# HOW WE DECIDE

Books by Jonah Lehrer

*Proust Was a Neuroscientist*

*How We Decide*

# How We
# DECIDE

Jonah Lehrer

HOUGHTON MIFFLIN HARCOURT

BOSTON · NEW YORK 2009

For information about permission to reproduce selections from
this book, write to Permissions, Houghton Mifflin Harcourt
Publishing Company, 215 Park Avenue South,
New York, New York 10003.

www.hmhbooks.com

*Library of Congress Cataloging-in-Publication Data*

Lehrer, Jonah.
How we decide / Jonah Lehrer.
p.   cm.
Includes bibliographical references and index.
ISBN 978-0-618-62011-1
1. Decision making. I. Title.
BF448.L45 2009
153.8'3—dc22   2008036769

Printed in the United States of America

Book design by Victoria Hartman

DOC 10 9 8 7 6 5 4 3 2 1

To my siblings,
*Eli, Rachel, and Leah*

# Contents

Who knows what I want to do? Who knows what anyone wants to do? How can you be sure about something like that? Isn't it all a question of brain chemistry, signals going back and forth, electrical energy in the cortex? How do you know whether something is really what you want to do or just some kind of nerve impulse in the brain? Some minor little activity takes place somewhere in this unimportant place in one of the brain hemispheres and suddenly I want to go to Montana or I don't want to go to Montana.

—DON DELILLO, *White Noise*

# Introduction

I was flying a Boeing 737 into Tokyo Narita International Airport when the left engine caught on fire. We were at seven thousand feet, with the runway dead ahead and the skyscrapers shimmering in the distance. Within seconds, bells and horns were blaring inside the cockpit, warning me of multiple system failures. Red lights flashed all over the place. I tried to suppress my panic by focusing on the automated engine-fire checklist, which told me to cut off fuel and power to the affected areas. Then the plane began a steep bank. The evening sky turned sideways. I struggled to steer the plane straight.

But I couldn't. The plane was impossible to fly. It swayed one way, I tried to pull her back to center, and then it swayed the other way. It was like wrestling with the atmosphere. Suddenly, I felt the shudder of a stall: the air was moving too slowly over the wings. The metal frame started to shriek and groan, the awful sound of steel giving way to physics. If I didn't find a way to increase my speed immediately, the plane would quickly surrender to the downward tug of gravity and I'd plunge into the city below.

I didn't know what to do. If I increased the throttle, I might be able to gain altitude and speed, and then I could circle around the runway and try to stabilize the plane. But could my only remaining engine handle the climb by itself? Or would it fail under the strain?

The other option was to steepen my descent in a desperate attempt to pick up speed; I'd fake a nosedive in order to avoid a real one. The downward momentum might let me avert the stall and steer the plane. Of course, I might instead be accelerating toward disaster. If I couldn't regain control, then the plane would fall into what pilots call a graveyard spiral. The g force would become so intense that the plane would disintegrate before it even hit the ground.

It was a hellish moment of indecision. Nervous sweat stung my eyes. My hands quivered with fear. I felt the blood pulse in my temples. I tried to think, but there wasn't time. The stall was getting worse. If I didn't act at that moment, the plane would fall out of the sky.

That's when I made up my mind: I would save the plane by taking her down. I tilted the yoke forward and prayed for speed. Immediately, I started to go faster. The problem was that I was heading straight into a suburb of Tokyo. But as my altimeter wound toward zero, the extra velocity kicked in and allowed me to steer. For the first time since the engine had caught fire, I could keep the plane on a steady course. I was still dropping like a stone, but at least I was flying in a straight line. I waited until the plane had sunk below two thousand feet and then pulled back on the yoke and advanced the throttle. The ride was sickeningly rough, but my descent remained on target. I lowered the landing gear and focused on keeping the plane under control, with the lights of the runway in the center of the windshield. My copilot called out the altitude: "One hundred feet! Fifty! Twenty!" Right before we hit the ground, I made a final plea for the center and waited for the comforting speed bump of solid earth. It was an ugly landing—I had to slam on the brakes and swerve at high speed—but we made it down safely.

It was only when the plane was parked at the airport gate that I noticed the pixels. I had been staring at a wraparound television screen, not looking through a cockpit window. The landscape below was just a quilt of satellite imagery. Although my hands were still shaking, nothing had really been at stake. There were no passengers sitting in the cabin. The Boeing 737 was just an artificial reality generated by a sixteen-million-dollar CAE Tropos 5000 flight simulator in a cavernous industrial hangar outside Montreal. My flight instructor had pressed a button and triggered the engine fire. (He'd also made my life more difficult by adding some fierce crosswinds.) But the flight had felt real. By the time the ride was over, my veins were full of adrenaline. A part of my brain was still convinced that I'd almost crashed into the city of Tokyo.

The virtue of a flight simulator is that you can investigate your own decisions. Had I been right to continue the descent? Or should I have tried to regain altitude? Would that have given me a smoother, safer landing? I wanted to know, so I asked the instructor if I could redo the simulated scenario and once again try to land without an engine. He flicked a few switches, and, before my heartbeat could recover, the 737 was reincarnated on the runway. I heard the voice of air-traffic control crackle on the radio, clearing me for takeoff. I advanced the throttle and sped down the tarmac. Everything went faster and faster until the aerodynamics took over and I was in the quiet of the evening's blue sky.

We climbed to ten thousand feet. I was just beginning to enjoy the tranquil view of Tokyo Bay when air-traffic control told me to prepare for landing. The scenario repeated itself like a familiar horror movie. I saw the same skyscrapers in the distance and flew through the same low clouds. I traced the same route across the same suburbs. I descended to nine thousand feet, then eight thousand, then seven thousand. And then it happened. The left engine erupted in flames. Once again, I struggled to keep the plane steady. Once again, there was the shudder warning me of a stall. This time, though, I aimed for the heavens. I increased the throttle, tilted the plane up, and carefully watched the readout from my one remaining engine. It soon became clear that I couldn't climb. There just wasn't enough engine power. The shudder spread across the skeleton of the plane. I heard the sickening sound of wings losing flight, a low resonant drone that filled the cockpit. The plane plunged left. A female voice calmly narrated the disaster, telling me what I already knew: I was falling out of the sky. The last thing I saw was a blink of city lights, just above the horizon. The screen froze when I hit the ground.

In the end, the difference between my landing the plane in one piece and my dying in a fiery crash came down to a single decision made in the panicked moments after the engine fire. It had all happened so fast, and I couldn't help but think about the lives that would have been at stake had this been a real flight. One decision led to a safe landing; the other to a fatal stall.

This book is about how we make decisions. It is about what happened inside my brain after the engine fire. It is about how the human mind—the most complicated object in the known universe—chooses what to do. It's about airplane pilots, NFL quarterbacks, television directors, poker players, professional investors, and serial killers, and the decisions they make every day. From the perspective of the brain, there's a thin line between a good decision and a bad decision, between trying to descend and trying to gain altitude. This book is about that line.

As long as people have made decisions, they've thought about how they make decisions. For centuries, they constructed elaborate theories on decision-making by observing human behavior from the outside. Since the mind was inaccessible—the brain was just a black box—these thinkers were forced to rely on un- testable assumptions about what was actually happening inside the head.

Ever since the ancient Greeks, these assumptions have re- volved around a single theme: humans are rational. When we make decisions, we are supposed to consciously analyze the al- ternatives and carefully weigh the pros and cons. In other words, we are deliberate and logical creatures. This simple idea under- lies the philosophy of Plato and Descartes; it forms the foun- dation of modern economics; it drove decades of research in cognitive science. Over time, our rationality came to define us. It was, simply put, what made us human.

There's only one problem with this assumption of human ra- tionality: it's wrong. It's not how the brain works. Look, for ex- ample, at my decisions in the cockpit. They were made in the heat of the moment, a visceral reaction to difficult events. I didn't carefully reflect on the best course of action or contemplate the aerodynamics of an engine fire. I couldn't reason my way to safety.

So how did I make a decision? What factors influenced my choices after the engine fire? For the first time in human history, these questions can be answered. We can look inside the brain and see how humans think: the black box has been broken open. It turns out that we weren't designed to be rational creatures. In- stead, the mind is composed of a messy network of different areas, many of which are involved with the production of emo- tion. Whenever someone makes a decision, the brain is awash in feeling, driven by its inexplicable passions. Even when a person tries to be reasonable and restrained, these emotional impulses secretly influence judgment. When I was in the cockpit desper- ately trying to figure out how to save my life—and the lives of thousands of Japanese suburbanites—these emotions drove the patterns of mental activity that made me crash and helped me land.

But this doesn't mean that our brains come preprogrammed for good decision-making. Despite the claims of many self-help books, intuition isn't a miraculous cure-all. Sometimes feelings can lead us astray and cause us to make all sorts of predictable mistakes. The human brain has a big cortex for a reason.

The simple truth of the matter is that making good decisions requires us to use both sides of the mind. For too long, we've treated human nature as an either/or situation. We are either rational or irrational. We either rely on statistics or trust our gut instincts. There's Apollonian logic versus Dionysian feeling; the id against the ego; the reptilian brain fighting the frontal lobes.

Not only are these dichotomies false, they're destructive. There is no universal solution to the problem of decision-making. The real world is just too complex. As a result, natural selection endowed us with a brain that is enthusiastically pluralist. Sometimes we need to reason through our options and carefully analyze the possibilities. And sometimes we need to listen to our emotions. The secret is knowing when to use these different styles of thought. We always need to be thinking about how we think.

This is what pilots learn in the flight simulator. The benefit of experiencing various cockpit scenarios—like an engine fire over Tokyo or a blizzard in Topeka—is that pilots develop better senses of which modes of thought to lean on in particular situations. "We never want pilots to act without thinking," says Jeff Roberts, the president of civil training at CAE, the largest manufacturer of flight simulators. "Pilots aren't robots, and that's a good thing. But we do want them to make decisions that rely on the wealth of judgment they've built up over time. You always need to think, but sometimes your feelings can help you think. A good pilot knows how to use his head."

At first, it might seem a little strange to look at decisions from the vantage point of the mind's inner workings. We aren't used to understanding choices in terms of competing brain regions or the firing rates of neurons. And yet, this new way of knowing ourselves—trying to understand human behavior from the inside—reveals many surprising things. In this book, you will learn how those three pounds of flesh inside the skull determine all of your decisions, from the most mundane choices in the supermarket to the weightiest of moral dilemmas. The mind inspires many myths—such as the fiction of pure rationality—but it's really just a powerful biological machine, complete with limitations and imperfections. Knowing how the machine works is useful knowledge, since it shows us how to get the most out of the machine.

But the brain doesn't exist in a vacuum; all decisions are made in the context of the real world. Herbert Simon, the Nobel Prize–winning psychologist, famously compared the human mind to a pair of scissors. One blade was the brain, he said, while the other blade was the specific environment in which the brain was operating.

If you want to understand the function of scissors, then you have to look at both blades simultaneously. To that end, we are going to venture out of the lab and into the real world so that we can see the scissors at work. I'll show you how the fluctuations of a few dopamine neurons saved a battleship during the Gulf War, and how the fevered activity of a single brain region led to the subprime housing bubble. We'll learn how firefighters handle dangerous blazes, and we'll visit the card tables of the World Series of Poker. We'll meet scientists who are using brain-imaging technology in order to understand how people make investment decisions and choose political candidates. I'll show you how some people are taking advantage of this new knowledge to make better television shows, win more football games, improve medical care, and enhance military intelligence. The goal of this book is to answer two questions that are of interest to just about everybody, from corporate CEOs to academic philosophers, from economists to airline pilots: How does the human mind make decisions? And how can we make those decisions better?

# 1

# The Quarterback in the Pocket

There is a minute and twenty-one seconds left on the clock in the 2002 Super Bowl, and the score is tied. The New England Patriots have the ball on their own 17-yard line. They are playing against the heavily favored St. Louis Rams. They have no time-outs left. Everyone assumes that the Patriots will kneel down and take the game into overtime. That, after all, is the prudent thing to do. "You don't want to have a turnover," says John Madden, one of the television broadcast's commentators. "They should just let time expire."

The game was never supposed to be this close. The Rams had been favored by fourteen points over the Patriots, which made this the most lopsided Super Bowl ever played. The potent Rams offense—nicknamed the "Greatest Show on Turf"—led the league in eighteen different statistical categories and outscored their opponents 503 to 273 during the regular season. Quarterback Kurt Warner was named the NFL's Most Valuable Player, and running back Marshall Faulk had won the NFL Offensive Player of the Year award. The Patriots, meanwhile, had been hamstrung by injuries, losing both Drew Bledsoe, their star quarterback, and Terry Glenn, their leading wide receiver. Everyone was expecting a rout.

But now, with just a minute remaining, Tom Brady—the second-string quarterback for the Patriots—has a chance to win the game. Over on the Patriots' sidelines, he huddles in conversation with Bill Belichick, the Patriots' head coach, and Charlie Weis, the offensive coordinator. "It was a ten-second conversation," Weis remembered later. "What we said is we would start the drive, and, if anything bad happened, we'd just run out the clock." The coaches were confident that their young quarterback wouldn't make a mistake.

Brady jogs back to his teammates on the field. You can see through his facemask that he's smiling, and it's not a nervous smile. It's a confident smile. There are seventy thousand spectators inside the Superdome, and most of them are rooting for the Rams, but Brady doesn't seem to notice. After a short huddle, the Patriots clap their hands in unison and saunter toward the line of scrimmage.

Tom Brady wasn't supposed to be here. He was the 199th pick in the 2000 draft. Although Brady had broken passing records at the University of Michigan, most team scouts thought he was too fragile to play with the big boys. The predraft report on Brady by *Pro Football Weekly* summarized the conventional wisdom: "Poor build. Very skinny and narrow. Ended the '99 season weighing 195 pounds, and still looks like a rail at 211. Lacks great physical stature and strength. Can get pushed down more easily than you'd like." The report devoted only a few words to Brady's positive attribute: "decision-making."

Belichick was one of the few coaches who had grasped Brady's potential. "Our vision wasn't that Tom was our franchise quarterback," Belichick said later, "but that Tom had been in situations—both in playing-time and game-management situations, tight games against good competition—and he'd handled all of them pretty well." Brady, in other words, had poise. He didn't choke under pressure. When the game was on the line, he found the open man.

Now Brady is in the spotlight, standing all by himself in the shotgun formation. His decision-making skills are about to be put to the test. He yells an audible to his tight end, then turns and yells at his wide receivers. The ball is snapped. Brady drops back, looks upfield, and understands instantly that the Rams have fallen into a tight zone coverage. They know the Patriots are going to pass; the cornerbacks are looking for an interception. Brady's primary target is covered, so he looks to his next target; he's also covered. Brady avoids the outstretched arm of a Ram defensive lineman, steps forward, and makes a short pass to his third target, the running back J. R. Redmond. It's a gain of five.

The next two plays unfold in the same way. Brady reads the Ram defense and calls out a series of coded commands: "White twenty! Ninety-six is the Mike! Omaha go!" These instructions tell the offensive linemen which linebackers to block and also serve as guides for the wide receivers, whose pass routes depend on the formation of the defense. After the play begins, Brady settles into the pocket, checks off his targets, and wisely settles for the safest option, which is a short pass in the flat. He doesn't force the ball into tight coverage. He's taking what the defense is giving him. The chains are moved, but the Patriots are running out of time.

It's now first and ten on the New England 41-yard line. Twenty-nine seconds remain in the game. Brady knows that he's got two, maybe three plays left. He has to move the ball another thirty yards just to get into field-goal range. The commentators sound like they're preparing for overtime, but the Patriots still think they can score. Brady settles into the shotgun. His eyes pan across the defense. He sees the linebackers edging a little closer to the line of scrimmage. Brady yells out the snap count, sends a man in motion, and then the ball is in his hands. He drops back and notices that only three defensive linemen are rushing him. The fourth is trying to cut off the short pass. Brady looks to his right. The receiver is covered. He looks to his left. Nobody's open. He looks at the center of the field. Troy Brown, a Patriots' wide receiver, is trying to find a plane of unoccupied space, a gap between the linebackers and the cornerbacks. Brady watches him clear the defenders and then fires a bullet fourteen yards downfield. Brown catches the ball in stride and runs for another nine yards before being pushed out-of-bounds. The ball is now thirty-six yards from the end zone, which is just within field-goal range. The Rams fans have gone silent.

With twelve seconds remaining, the Patriots' special-teams unit is brought onto the field. Adam Vinatieri steps into the forty-eight-yard kick. The ball sails straight between the pylons. The clock says triple zero. The Patriots have just won the Super Bowl. It's the greatest upset in NFL history.

# 1

The quick decisions made by a quarterback on a football field provide a window into the inner workings of the brain. In the space of a few frenetic seconds, before a linebacker crushes him into the ground, an NFL quarterback has to make a series of hard choices. The pocket is collapsing around him—the pocket begins to collapse before it exists—but he can't flinch or wince. His eyes must stay focused downfield, looking for some meaningful sign amid the action, an open man on a crowded field. Throwing the ball is the easy part.

These passing decisions happen so fast they don't even seem like decisions. We are used to seeing football on television, captured by the cameras far above the grassy stage. From this distant perspective, the players appear to be moving in some sort of violent ballet; the sport looks exquisitely choreographed. You can see the receivers spread the zone and watch the pocket slowly disintegrate. It's easy to detect the weak spots of the defense and find the target with man-on-man coverage. You can tell which linebackers bought the play-action fake and see the cornerback racing in on the blitz. When you watch the game from this omniscient angle—coaches call it "the eye in the sky"—it appears as if the quarterback is simply following orders, as if he knows where he is going to throw the ball before the play begins.

But this view of the game is deeply misleading. After the ball is snapped, the ordered sequence of neat X's and O's that fill the spiral-bound playbook degenerates into a street brawl. There's a symphony of grunts and groans and the meaty echoes of fat men hitting hard ground. Receivers get pushed off their routes, passing angles get cut off, and inside blitzes derail the best intentions. The offensive line is an unpredictable wrestling match. Before the quarterback can make an effective decision, he needs to assimilate all of this new information and be aware of the approximate location of every player on the field.

The savage chaos of the game, the way every play is a mixture of careful planning and risky improvisation, is what makes the job of an NFL quarterback so difficult. Even while he's immersed in the violence—the defensive line clawing at his body—the quarterback has to stand still and concentrate. He needs to look past the mayhem and make sense of all the moving bodies. Where is his receiver going? Will the safety break toward the ball? Is the linebacker going to drop back into coverage? Did his tight end pick up the blitz? Before a pass can be thrown—before the open man can be found—all of these questions need to be answered. Each pass is really a guess, a hypothesis launched into the air, but the best quarterbacks find ways to make better guesses. What separates Tom Brady and Joe Montana and Peyton Manning and John Elway and the other great quarterbacks of the modern NFL era from the rest is their ability to find the right receiver at the right time. (The Patriots like to pass out of a five-wide formation, which means that Brady often checks off five different receivers before he decides where to throw the ball.) No other team sport is so dependent on the judgment of a single player.

NFL scouts take the decision-making skills of quarterbacks very seriously. The league requires that every player in the draft take the Wonderlic intelligence test, which is essentially a shorter version of the standard IQ test. The test is twelve minutes long and consists of fifty questions that get progressively harder as the test goes along. Here's an example of an easy Wonderlic question:

"Paper sells for 21 cents per pad. What will four pads cost?"

And here's a hard Wonderlic question:

"Three individuals form a partnership and agree to divide the profits equally. X invests $9,000, Y invests $7,000, Z invests $4,000. If the profits are $4,800, how much less does X receive than if the profits were divided in proportion to the amount invested?"

The underlying thesis of the Wonderlic test is that players who are better at math and logic problems will make better decisions in the pocket. At first glance, this seems like a reasonable assumption. No other position in sports requires such extreme cognitive talents. Successful quarterbacks need to memorize hundreds of offensive plays and dozens of different defensive formations. They need to spend hours studying game tape of their opponents and be able to put that knowledge to use on the field. In many instances, quarterbacks are even responsible for changing plays at the line of scrimmage. They are like coaches with shoulder pads.

As a result, an NFL team starts to get nervous when a quarterback's score on the Wonderlic test is too far below the average for the position. For quarterbacks, the average is 25. (In comparison, the average score for computer programmers is 28. Janitors, on average, score 15, as do running backs.) Vince Young, the star quarterback from the University of Texas, reportedly scored a 6 on the test, which led many teams to publicly question his ability to succeed in the NFL.

But Young ended up excelling in the pros. And he isn't the only quarterback who achieved success despite a poor Wonderlic score. Dan Marino scored 14. Brett Favre's Wonderlic score was 22, and Randall Cunningham and Terry Bradshaw both scored 15. All of these quarterbacks have been or will be inducted into the Hall of Fame. (In recent years, Favre has surpassed many of the passing records once held by Marino, such as most passing yards and touchdowns in a career.) Furthermore, several quarterbacks with unusually high Wonderlic scores—players like Alex Smith and Matt Leinart, who both scored above 35 on the test and were top-ten picks in the 2005 NFL draft—have struggled in the NFL, largely because they make poor decisions on the field.

The reason there is virtually no correlation between the results of the Wonderlic and the success of quarterbacks in the NFL is that finding the open man involves a very different set of decision-making skills than solving an algebra problem. While quarterbacks need to grapple with complexity—the typical offensive playbook is several inches thick—they don't make sense of the football field the way they make sense of questions on a multiple-choice exam. The Wonderlic measures a specific kind of thought process, but the best quarterbacks don't think in the pocket. There isn't time.

Take that pass to Troy Brown. Brady's decision depended on a long list of variables. He needed to know that the linebacker wouldn't fall back into coverage and that there were no cornerbacks in the area waiting for an interception. After that, he had to calculate the ideal place to hit Brown with the ball so that

Brown would have plenty of room to run after the catch. Then he needed to figure out how to make a throw without hitting the defensive lineman blocking his passing lane. If Brady were forced to consciously analyze this decision—if he treated it like a question on the Wonderlic test—then every pass would require a lot of complicated trigonometry as he computed his passing angles on the plane of the football field. But how can you contemplate the math when five angry linemen are running straight at you? The answer is simple: you can't. If a quarterback hesitates for even a split second, he is going to get sacked.

So how does a quarterback do it? How does he make a decision? It's like asking a baseball player why he decided to swing the bat at a particular pitch: the velocity of the game makes thought impossible. Brady can afford to give each receiver only a split second of attention before he has to move on to the next. As soon as he glances at a body in motion, he must immediately decide if that body will be open a few seconds in the future. As a result, a quarterback is forced to evaluate each of his passing alternatives without knowing how he's evaluating them. Brady chooses a target without understanding why exactly he's settled on that target. Did he pass to Troy Brown with twenty-nine seconds remaining in the Super Bowl because the middle linebacker had ceded too much space, or because the cornerbacks were following the other receivers downfield and leaving a small gap in the center of the field? Or did Brady settle on Brown because all the other passing options were tightly covered, and he knew that he needed a long completion? The quarterback can't answer these questions. It's as if his mind is making decisions without him. Even quarterbacks are mystified by their talents. "I don't know how I know where to pass," Brady says. "There are no firm rules. You just feel like you're going to the right place . . . And that's where I throw it."

# 2

The mystery of how we make decisions—how Tom Brady chooses where to throw the ball—is one of the oldest mysteries of the mind. Even though we are defined by our decisions, we are often completely unaware of what's happening inside our heads during the decision-making process. You can't explain why you bought the box of Honey Nut Cheerios, or stopped at the yellow traffic light, or threw the football to Troy Brown. On the evaluation sheets of NFL scouts, *decision-making* is listed in the category Intangibles. It's one of the most important qualities in a quarterback, and yet nobody knows what it is.

The opaque nature of this mental process has led to a surfeit of theorizing. The most popular theory frames decision-making in epic terms, as a pitched battle between reason and emotion, with reason often triumphing. According to this classic script, what separates us from animals is the godly gift of rationality. When we are deciding what to do, we are able to ignore our feelings and carefully think through the problem. A quarterback, for instance, is supposed to choose a receiver by calmly contemplating all of the information on the field, translating the helter-skelter of the pass play into a series of discrete math problems. A more rational quarterback, with a higher Wonderlic score, should be a better quarterback. This ability to analyze the facts—to transcend our feelings, instincts, and impulses—is often seen as the defining element of human nature.

Plato, as usual, was there first. He liked to imagine the mind as a chariot pulled by two horses. The rational brain, he said, is the charioteer; it holds the reins and decides where the horses run. If the horses get out of control, the charioteer just needs to take out his whip and reassert authority. One of the horses is well bred and well behaved, but even the best charioteer has difficulty controlling the other horse. "He is of an ignoble breed,"

Plato wrote. "He has a short bull-neck, a pug nose, black skin, and bloodshot white eyes; companion to wild boasts and indecency, he is shaggy around the ears—deaf as a post—and just barely yields to horsewhip and goad combined." According to Plato, this obstinate horse represents negative, destructive emotions. The job of the charioteer is to keep the dark horse from running wild and to keep both horses moving forward.

With that single metaphor, Plato divided the mind into two separate spheres. The soul was seen as conflicted, torn between reason and emotion. When the driver and horses wanted different things, Plato said, it was essential to listen to the driver. "If the better elements of the mind which lead to order and philosophy prevail," he wrote, "then we can lead a life here in happiness and harmony, masters of ourselves." The alternative, he warned, was a life governed by impulsive emotions. If we follow the horses, we will be led like a "fool into the world below."

This division of the mind is one of Plato's most enduring themes, an idea enshrined in Western culture. On the one hand, humans are part animal, primitive beasts stuffed full of primitive desires. And yet, humans are also capable of reason and foresight, blessed with the divine gift of rationality. The Roman poet Ovid, writing in *Metamorphoses* a few centuries after Plato, captured this psychology in a few short sentences. Medea has fallen in love with Jason—she was literally struck by Eros's arrow—but this love conflicts with her duty to her father. "I am dragged along by a strange new force," she laments. "Desire and reason are pulling in different directions. I see the right way and approve it, but follow the wrong."

René Descartes, the most influential philosopher of the Enlightenment, agreed with this ancient critique of feeling. Descartes divided our being into two distinct substances: a holy soul capable of reason, and a fleshy body full of "mechanical passions." What Descartes wanted to do was purge the human intellect of its falsehoods, to get beyond the illogical beliefs of the past. In his seminal work, the awkwardly titled *Discourse on the Method for Properly Conducting Reason and Searching for Truth,* Descartes tried to provide an example of rationality in pure form. His goal was to lead humanity out of the cave, to reveal the "clear and distinct" principles that our emotions and intuitions obscure.

The Cartesian faith in reason became a founding principle of modern philosophy. Rationality was like a scalpel, able to dissect reality into its necessary parts. Emotions, on the other hand, were crude and primitive. Over time, a variety of influential thinkers tried to translate this binary psychology into practical terms. Francis Bacon and Auguste Comte wanted to reorganize society so that it reflected "rational science"; Thomas Jefferson hoped that the "American experiment would prove that men can be governed by reason and reason alone"; Immanuel Kant came up with the concept of the categorical imperative so that morality *was* rationality. At the height of the French Revolution, a group of radicals founded the Cult of Reason and turned several Parisian cathedrals into temples of rationality. There were no temples dedicated to emotion.

The twentieth-century version of the Platonic metaphor was put forth by Sigmund Freud. Although Freud liked to say that he spent his life destroying illusions, his basic view of the mind differed little from Plato's. In his "speculative science," Freud imagined the human mind as divided into a series of conflicting parts. (Conflict was important to Freud, since it helped explain neuroses.) At the center of the mind was the id, a factory of crude desires. Above that was the ego, which represented the conscious self and the rational brain. It was the job of the ego to restrain the id, channeling its animal emotions in socially acceptable ways. "One might compare the relations of the ego to the id with that between a rider and his horse," Freud wrote in a direct allusion to Plato. "The horse provides the locomotive energy, and the rider has the prerogative of determining the goal and of guiding the movements of his powerful mount towards it."

The purpose of Freudian psychoanalysis was to fortify the ego, to build up the strength needed to control the impulses of the id. In other words, Freud tried to teach his patients how to hold back their horses. He believed that most mental disorders, from hysteria to narcissism, were due to the effects of unrestrained feelings. In later years, Freud would turn this Platonic vision into a theory of everything. He saw civilization, or *kultur,* as the individual mind writ large. "The events of human history," Freud wrote, "are only the reflections of the dynamic conflicts among the id and ego, which psychoanalysis studies in the individual—the same events on a wider stage." According to Freud, the survival of modern society depended on people sacrificing the emotional desires of their ids—what he termed the pleasure principle—for the sake of the greater good. The possibility of human reason was the only thing that kept civilization from descending into barbarism. As Goya put it, "The sleep of reason produces monsters."

Over time, Freudian psychology lost its scientific credibility. Discussions of the id, ego, and Oedipus complex were replaced by references to specific areas in the brain; Viennese theory gave way to increasingly exact anatomical maps of the cortex. The metaphor of the Platonic chariot seemed woefully obsolete.

But modern science soon hit on a new metaphor: the mind was a computer. According to cognitive psychology, each of us was a set of software programs running on three pounds of neural hardware. While this computer metaphor helped stimulate some important scientific breakthroughs—it led to the birth of artificial intelligence, among other things—it was also misleading, at least in one crucial respect. The problem with seeing the mind as a computer is that computers don't have feelings. Because emotions couldn't be reduced to bits of information or the logical structures of programming language, scientists tended to ignore them. "Cognitive psychologists subscribed to this false ideal of rational, logical thought, and so we diminished the importance of everything else," says Marvin Minsky, a professor at MIT and a pioneer of artificial intelligence. When cognitive psychologists did think about emotion, they tended to reinforce the Platonic divide: feelings interfered with cognition. They were the antagonists of rationality, and they messed up the machine. That was the version of the mind put forth by modern science.

The simple idea connecting Plato's philosophy to cognitive psychology is the privileging of reason over emotion. It's easy to understand why this vision has endured for so long. It raises *Homo sapiens* above every other animal: the human mind is a rational computer, a peerless processor of information. Yet it also helps explain away our flaws: because each of us is still part animal, the faculty of reason is forced to compete with primitive emotions. The charioteer must control those wild horses.

This theory of human nature comes with a corollary: if our feelings keep us from making rational decisions, then surely we'd be better off without any feelings at all. Plato, for example, couldn't help but imagine a utopia in which reason determined everything. Such a mythical society—a republic of pure reason—has been dreamed of by philosophers ever since.

But this classical theory is founded upon a crucial mistake. For too long, people have disparaged the emotional brain, blaming our feelings for all of our mistakes. The truth is far more interesting. What we discover when we look at the brain is that the horses and the charioteer depend upon each other. If it weren't for our emotions, reason wouldn't exist at all.

# 3

In 1982, a patient named Elliot walked into the office of neurologist Antonio Damasio. A few months earlier, a small tumor had been cut out of Elliot's cortex, near the frontal lobe of his brain. Before the surgery, Elliot had been a model father and husband. He'd held down an important management job in a large corporation and was active in his local church. But the operation changed everything. Although Elliot's IQ had stayed the same—he still tested in the 97th percentile—he now exhibited one psychological flaw: he was incapable of making a decision.

This dysfunction made normal life impossible. Routine tasks that should have taken ten minutes now required several hours. Elliot endlessly deliberated over irrelevant details, like whether to use a blue or black pen, what radio station to listen to, and where to park his car. When he chose where to eat lunch, Elliot carefully considered each restaurant's menu, seating plan, and lighting scheme, and then drove to each place to see how busy it was. But all this analysis was for naught: Elliot still couldn't decide where to eat. His indecision was pathological.

Before long, Elliot was fired from his job. That's when things really began to fall apart. He started a series of new businesses, but they all failed. He was taken in by a con man and was forced into bankruptcy. His wife divorced him. The IRS began an investigation. He moved back in with his parents. As Damasio put it, "Elliot emerged as a man with a normal intellect who was unable to decide properly, especially when the decision involved personal or social matters."

But why was Elliot suddenly incapable of making good decisions? What had happened to his brain? Damasio's first insight occurred while talking to Elliot about the tragic turn his life had taken. "He was always controlled," Damasio remembers, "always describing scenes as a dispassionate, uninvolved spectator. Nowhere was there a sense of his own suffering, even though he was the protagonist . . . I never saw a tinge of emotion in my many hours of conversation with him: no sadness, no impatience, no frustration." Elliot's friends and family confirmed Damasio's observations: ever since his surgery, he'd seemed strangely devoid of emotion, numb to the tragic turn his own life had taken.

To test this diagnosis, Damasio hooked Elliot to a machine that measured the activity of the sweat glands in his palms. (When a person experiences strong emotions, the skin is literally aroused and the hands start to perspire. Lie detectors operate on the basis of this principle.) Damasio then showed Elliot various photographs that normally triggered an immediate emotional response: a severed foot, a naked woman, a house on fire, a handgun. The results were clear: Elliot felt nothing. No matter how grotesque or aggressive the picture, his palms never got sweaty. He had the emotional life of a mannequin.

This was a completely unexpected discovery. At the time, neuroscience assumed that human emotions were *irrational*. A person without any emotions — in other words, someone like Elliot — should therefore make better decisions. His cognition should be uncorrupted. The charioteer should have complete control.

What, then, had happened to Elliot? Why couldn't he lead a normal life? To Damasio, Elliot's pathology suggested that emotions are a crucial part of the decision-making process. When we are cut off from our feelings, the most banal decisions became impossible. A brain that can't feel can't make up its mind.


AFTER INTERVIEWING ELLIOT, Damasio began studying other patients with similar patterns of brain damage. These patients all appeared intelligent and showed no deficits on any conventional cognitive tests. And yet they all suffered from the same profound flaw: because they didn't experience emotion, they had tremendous difficulty making any decisions. In *Descartes' Error,*

Damasio described what it was like trying to set up an appointment with one of these emotionless patients:

> I suggested two alternative dates, both in the coming month and just a few days apart from each other. The patient pulled out his appointment book and began consulting the calendar. The behavior that ensued, which was witnessed by several investigators, was remarkable. For the better part of a half hour, the patient enumerated reasons for and against each of the two dates: previous engagements, proximity to other engagements, possible meteorological conditions, virtually anything that one could reasonably think about concerning a simple date. . . . He was now walking us through a tiresome cost-benefit analysis, an endless outlining and fruitless comparison of options and possible consequences. It took enormous discipline to listen to all of this without pounding on the table and telling him to stop.

Based on these patients, Damasio began compiling a map of feeling, locating the specific brain regions responsible for generating emotions. Although many different cortical areas contribute to this process, one part of the brain seemed particularly important: a small circuit of tissue called the orbitofrontal cortex, which sits just behind the eyes, in the underbelly of the frontal lobe. (*Orbit* is Latin for "eye socket.") If this fragile fold of cells is damaged by a malignant tumor or a hemorrhaging artery, the tragic result is always the same. At first, everything seems normal, and after the tumor is removed or the bleeding is stopped, the patient is sent home. A full recovery is forecast. But then little things start to go awry. The patient begins to seem remote, cold, distant. This previously responsible person suddenly starts doing irresponsible things. The mundane choices of everyday life become excruciatingly difficult. It's as if his very personality—the collection of wants and desires that defined him as an individual—had been systematically erased. His loved ones say it's like living with a stranger, only this stranger has no scruples.

The crucial importance of our emotions—the fact that we can't make decisions without them—contradicts the conventional view of human nature, with its ancient philosophical roots. For most of the twentieth century, the ideal of rationality was supported by scientific descriptions of human anatomy. The brain was envisioned as consisting of four separate layers, stacked in ascending order of complexity. (The cortex was like an archaeological site: the deeper you dug, the farther back in time you traveled.) Scientists explained the anatomy of the human brain in this manner: At its bottom was the brain stem, which governed the most basic bodily functions. It controlled heartbeat, breathing, and body temperature. Above that was the diencephalon, which regulated hunger pangs and sleep cycles. Then came the limbic region, which generated animal emotions. It was the source of lust, violence, and impulsive behavior. (Human beings shared these three brain layers with every other mammal.) Finally, there was the magnificent frontal cortex—the masterpiece of evolution—which was responsible for reason, intelligence, and morality. These convolutions of gray matter allowed each of us to resist urges and suppress emotions. In other words, the rational fourth layer of the brain allowed us to ignore the first three layers. We were the only species able to rebel against primitive feelings and make decisions that were dispassionate and deliberate.

But this anatomical narrative is *false*. The expansion of the frontal cortex during human evolution did not turn us into purely rational creatures, able to ignore our impulses. In fact, neuroscience now knows that the opposite is true: a significant part of our frontal cortex is involved with emotion. David Hume, the eighteenth-century Scottish philosopher who delighted in heretical ideas, was right when he declared that reason was "the slave of the passions."

How does this emotional brain system work? The orbitofrontal cortex (OFC), the part of the brain that Elliot was missing, is responsible for integrating visceral emotions into the decision-making process. It connects the feelings generated by the "primitive" brain—areas like the brain stem and the amygdala, which is in the limbic system—to the stream of conscious thought. When a person is drawn to a specific receiver, or a certain entrée on the menu, or a particular romantic prospect, the mind is trying to tell him that he should choose that option. It has already assessed the alternatives—this analysis takes place outside of conscious awareness—and converted that assessment into a positive emotion. And when he sees a receiver who's tightly covered, or smells a food he doesn't like, or glimpses an ex-girlfriend, it is the OFC that makes him want to get away. (*Emotion* and *motivation* share the same Latin root, *movere*, which means "to move.") The world is full of things, and it is our feelings that help us choose among them.

When this neural connection is severed—when our OFCs can't comprehend our own emotions—we lose access to the wealth of opinions that we normally rely on. All of a sudden, you no longer know what to think about the receiver running a short post pattern or whether it's a good idea to order the cheeseburger for lunch. The end result is that it's impossible to make decent decisions. This is why the OFC is one of the few cortical regions that are markedly larger in humans than they are in other primates. While Plato and Freud would have guessed that the job of the OFC was to *protect* us from our emotions, to fortify reason against feeling, its actual function is precisely the opposite. From the perspective of the human brain, *Homo sapiens* is the most emotional animal of all.

4

It's not easy making a daytime soap opera. The demands of the form are grueling: a new episode has to be filmed nearly every single day. No other type of popular entertainment churns out so much material in so short a time. New plot twists have to be dreamed up, new scripts have to be written, actors need to rehearse, and every scene must be meticulously mapped out. Only then, once all that preparation is complete, are the cameras turned on. For most daytime soaps, it takes about twelve hours to film twenty-two minutes of television. This cycle is repeated five days a week.

Herb Stein has been directing *Days of Our Lives,* a soap opera on NBC, for twenty-five years. He's shot more than fifty thousand scenes and has cast hundreds of different actors. He's been nominated for eight daytime Emmys. Over the course of his long career, Stein has witnessed more scenes of melodrama—rapes, weddings, births, murders, confessions—than just about any other human being alive. He is, one might say, an *expert* on melodrama: how to write it, block it, film it, edit it, and produce it.

For Stein, the long road to daytime television began when he was a student at UCLA and read *The Oresteia,* the trilogy of classic Greek tragedies written by Aeschylus. It was the utter timelessness of the plays—their ability to speak to enduring human themes—that made him want to study theater. When Stein talks about drama—and it doesn't matter if he's talking about Aeschylus or *General Hospital*—he tends to sound like a literature professor. (He also looks like one, with his rumpled shirts and a few days' worth of salt-and-pepper stubble.) Stein talks in long, digressive monologues and finds grand ideas in the most unlikely plot lines. "Many of these classic plays have elements of the ridiculous," he says. "The plots are often completely implausible. That whole Oedipus thing? Totally absurd. And yet, when these stories are told well, you don't notice the absurdity. You're too busy paying attention to what's happening."

Soap operas work the same way. The key to being a successful soap opera director—and Stein is one of the most successful in the business—is telling the story so that people don't notice you're telling them a story. Everything has to feel sincere, even when what's happening onscreen is completely outlandish. This is much harder than it might seem. Let's say you're shooting a scene in which a woman is giving birth to fraternal twins fathered by two different men, both of whom are at the bedside with her. One of the fathers is the villain of the show: he impregnated the woman by raping her. The other father is the good guy, and the woman is deeply in love with him. However, if she doesn't marry her rapist, then members of her family will be killed. (This is an actual plot line from a recent *Days of Our Lives* episode.) The scene has several pages of intense dialogue, a few tears, and plenty of subtext. Stein has about an hour to shoot it, which forces him to make some crucial decisions on the fly. He has to figure out where each character should stand, how they all should move, what emotions they should convey, and how each of the four cameras should capture the action. Should they zoom in close, or get a reaction shot over the shoulder? How should the villain deliver his lines? These directorial decisions will determine whether or not the scene works. "You've really got to know how to milk the drama," Stein says. "Otherwise, it's just a bunch of people standing in a room, saying stupid stuff."

Although the scene has been mapped out in advance, Stein still needs to make many of these decisions in the midst of filming, while the actors are delivering their lines. Most of the fake rooms on the Burbank sound stage have only two flimsy walls, with one camera positioned on each side. An additional camera records the center of the scene. As soon as the assistant director yells out, "Action!" there is a frenzy of activity offstage as the cameras pivot and Stein snaps his fingers, pointing to indicate which camera he wants to capture the action for each specific part of the scene. (This makes it easier for the editor to assemble a working cut later.) During complicated scenes, such as that birth scene with the two fathers, Stein looks like an orchestra conductor: his arms are never still. He is constantly pointing at different cameras, crafting the scene in real time.

How does Stein make these directorial decisions? After all, he doesn't have the luxury of filming twenty different takes from twenty different angles. "Given the schedule [of a daytime soap opera]," Stein says, "there isn't time to be fiddling around with all the stuff that directors normally fiddle around with. You need to make the right decision the first time around." If a soap director makes a mistake while shooting, the scene can't be re-shot another day. When you're creating daytime television, you have only one day.

This relentless time pressure means that Stein can't afford to carefully think through all of his camera choices. He doesn't have time to be rational; he needs to react to the drama as it's unfolding. In that sense, he is like a quarterback in the pocket. "When you shoot as many scenes as I have," Stein says, "you just know how things should go. I can watch an actor say a single line and know immediately that we need to try it again. When we're filming a scene, it's all very instinctual. Even when we go in with a plan on how to shoot it, that plan will often change in the moment, depending on how it feels."

The reliance on instinct and "feel" is also a crucial part of the casting process. Soaps are continually bringing on new actors, in part because the longer actors are on the show, the higher their salaries are. (That's why established characters on *Days of Our Lives* are constantly being killed off. As Stein quips, "This isn't show *art*. It's show *business*.") For a soap opera, there are few decisions as crucial as casting. The size of the audience oscillates with the appeal of the actors, and a particularly appealing actor can create a spike in the ratings. "You are always looking for that person that people want to look at," Stein says. "And I don't just mean attractiveness. They've got to have *it,* and by *it,* I mean everything that you can't really put into words."

The question, of course, is how you identify *it*. When Stein first started directing television, he was overwhelmed by all the different variables involved in casting. First, he'd try to make sure that the person looked right for the part and could act in the soap style. Then Stein needed to consider how this actor would fit in with the rest of the cast. ("A lack of chemistry has ruined many a soap scene," he says.) Only after that was Stein able to think about whether or not the actor actually had talent. Would he deliver his lines with sincerity? Could he cry on demand? How many takes would he require before he got the scene right? "Given all of these factors," Stein says, "there can be a tendency to really outthink yourself, to talk yourself into choosing the wrong actor."

After directing daytime television for decades, however, Stein has learned how to trust his instincts, even if he can't always explain them. "It only takes me three to five seconds before I know if the person is right," he says. "A few words, a single gesture. That's all I need. And I've learned to always listen to that." Recently, the show put out a casting call for a male lead. The character was going to be the new villain on the show. Stein was up in his office, blocking a script, watching the auditions out of the corner of his eye. After a few hours of seeing dozens of different actors recite the exact same lines, Stein was getting bored and discouraged. "But that's when this one guy stood up," he says. "The actor didn't even know his lines because he had gotten the script late. I just saw him say a few words, and then I knew. He was unbelievably great. I couldn't explain why, but for me he completely stood out. What they say is true: you just get a feeling."

The mental process Stein is describing depends on his emotional brain. Those twinges of feeling that help him select the right camera and find the best actor are a distillation of all those details that he doesn't consciously perceive. "The conscious brain may get all the attention," says Joseph LeDoux, a neuroscientist at NYU. "But consciousness is a small part of what the brain does, and it's a slave to everything that works beneath it." According to LeDoux, much of what we "think" is really driven by our emotions. In this sense, every feeling is really a summary of data, a visceral response to all of the information that can't be accessed directly. While Stein's conscious brain was blocking the script, his unconscious supercomputer was processing all sorts of data. It then translated that data into vivid emotional signals that were detected by the OFC, allowing Stein to act upon these subliminal calculations. If Stein were missing his feelings—if he were like one of Damasio's patients—then he would be forced to carefully analyze every alternative, and that would take forever. His episodes would be constantly delayed and he would cast the wrong actors. Stein's insight is that his feelings are often an accurate shortcut, a concise expression of his decades' worth of experience. They already know how to shoot the scene.

WHY ARE OUR emotions so essential? How did they get so good at finding the open man and directing soap operas? The answer is rooted in evolution. It takes a long time to design a brain. The first clumps of networked neurons appeared more than five hundred million years ago. This was the first nervous system, although at that point it was really just a set of automatic reflexes. Over time, however, primitive brains grew increasingly complex. They expanded from a few thousand neurons in earthworms to a hundred billion connected cells in Old World primates. When *Homo sapiens* first appeared, about two hundred thousand years ago, the planet was already full of creatures with highly specialized brains. There were fish that could migrate across the ocean using magnetic fields, and birds that navigated by starlight, and insects that could smell food from a mile away. These cognitive feats were all byproducts of instincts that had been engineered by natural selection to perform specific tasks. What these animals couldn't do, however, was reflect on their own decisions. They couldn't plan out their days or use language to express their inner states. They weren't able to analyze complex phenomena or invent new tools. What couldn't be done automatically couldn't be done at all. The charioteer had yet to appear.

The evolution of the human brain changed everything. For the first time, there was an animal that could think about how it thought. We humans could contemplate our emotions and use words to dissect the world, parsing reality into neat chains of causation. We could accumulate knowledge and logically analyze problems. We could tell elaborate lies and make plans for the future. Sometimes, we could even follow our plans.

These new talents were incredibly useful. But they were also incredibly new. As a result, the parts of the human brain that make them possible—the ones that the driver of the chariot controls—suffer from the same problem that afflicts any new technology: they have lots of design flaws and software bugs. (The human brain is like a computer operating system that was rushed to market.) This is why a cheap calculator can do arithmetic better than a professional mathematician, why a mainframe computer can beat a grand master at chess, and why we so often confuse causation and correlation. When it comes to the new parts of the brain, evolution just hasn't had time to work out the kinks.

The emotional brain, however, has been exquisitely refined by evolution over the last several hundred million years. Its software code has been subjected to endless tests, so it can make fast decisions based on very little information. Look, for instance, at the mental process involved in hitting a baseball. The numbers make the task look impossible. A typical major-league pitch takes about 0.35 seconds to travel from the hand of the pitcher to home plate. (This is the average interval between human heartbeats.) Unfortunately for the batter, it takes about 0.25 seconds for his muscles to initiate a swing, leaving his brain a paltry one-tenth of a second to make up its mind on whether or not to do so. But even this estimate is too generous. It takes a few milliseconds for the visual information to travel from the retina to the visual cortex, so the batter really has fewer than five milliseconds to perceive the pitch and decide if he should swing. But people can't think this quickly; even under perfect conditions, it takes the brain about twenty milliseconds to respond to a sensory stimulus.

So how does a major-league baseball player manage to hit a fastball? The answer is that the brain begins collecting information about the pitch long before the ball leaves the pitcher's hand. As soon as the pitcher begins his wind-up, the batter automatically starts to pick up on "anticipatory clues" that help him winnow down the list of possibilities. A torqued wrist suggests a curveball, while an elbow fixed at a right angle means that a fastball is coming, straight over the plate. Two fingers on the seam might indicate a slider, and a ball gripped with the knuckles is a sure sign that a wavering knuckleball is on its way. The batters, of course, aren't consciously studying these signs; they can't tell you why they decided to swing at certain pitches. And yet, they are able to act based on this information. For instance, a study of expert cricket batters demonstrated that the players could accurately predict the speed and location of the ball based solely on a one-second video of the pitcher's wind-up. The well-trained brain knew exactly what details to look for. And then, once it perceived these details, it seamlessly converted them into an accurate set of feelings. For a hitter in the major leagues, a hanging curveball over the center of the plate just *feels* like a better pitch than a slider, low and away.

We take these automatic talents for granted precisely because they work so well. There's no robot that can hit a baseball or throw a football or ride a bicycle. No computer program can figure out which actor should play a villain or instantly recognize a familiar face. This is why when evolution was building the brain, it didn't bother to replace all of those emotional processes with new operations under explicit, conscious control. If something isn't broken, then natural selection isn't going to fix it. The mind is made out of used parts, engineered by a blind watchmaker. The result is that the uniquely human areas of the mind depend on the primitive mind underneath. The process of thinking requires feeling, for feelings are what let us understand all the information that we can't directly comprehend. Reason without emotion is impotent.

One of the first scientists to defend this view of decision-making was William James, the great American psychologist. In his seminal 1890 textbook *The Principles of Psychology*, James launched into a critique of the standard "rationalist" account of the human mind. "The facts of the case are really tolerably plain," James wrote. "Man has a far *greater* variety of impulses than any other lower animal." In other words, the Platonic view of decision-making, which idealized man as a purely rational animal defined "by the almost total absence of instincts," was utterly mistaken. James's real insight, however, was that these impulses weren't necessarily bad influences. In fact, he believed that "the preponderance of habits, instincts and emotions" in the human brain was an essential part of what made the brain so effective. According to James, the mind contained two distinct thinking systems, one that was rational and deliberate and another that was quick, effortless, and emotional. The key to making decisions, James said, was knowing when to rely on which system.

Just look at Tom Brady. It's his feelings that allow him to make quick passing decisions in the pocket. For Brady, the process probably works something like this: After the ball is snapped, he drops back and tries to make sense of the field. He begins going through his checklist of receivers. The primary target, a tight end running a short crossing pattern, is tightly covered. As a result, when Brady glances at the tight end, he automatically feels a slight twinge of fear, the sure sign of a risky pass. The presence of the linebacker has been translated into a negative emotion. Brady then proceeds to his secondary target, a wide receiver running a deep out. Unfortunately, this target is double-teamed by a cornerback and a safety. Once again, Brady experiences a negative feeling, an instant distillation of what's happening on the football field. A few seconds have now elapsed, and Brady can feel the pressure of the defensive line. His left tackle is being pushed backward; Brady knows that he's got to get rid of the ball soon or the game is going to end with a sack. He proceeds to his third target. Troy Brown is streaking across the center of the field, threading the seam between the linebackers and the cornerbacks. When Brady looks at this target, his usual fear is replaced by a subtle burst of positive emotion, the allure of a receiver without a nearby defender. He has found the open man. He lets the ball fly.

# 2

# The Predictions of Dopamine

In the early morning hours of February 24, 1991, the First and Second Marine divisions rolled north across the desert of Saudi Arabia. As they approached the unmarked border with Kuwait — the landscape was just an expanse of barren sand — the troops accelerated their pace. These Marines were the first Coalition forces to enter the country since it had been invaded by Iraq, more than eight months earlier. The outcome of Operation Desert Storm depended on their success. The Marines needed to liberate Kuwait, and they needed to do it in fewer than one hundred hours. If the Marines failed to overtake the Iraqi army quickly, they faced the prospect of urban warfare. The Iraqis were threatening to retreat into the streets of Kuwait City, and if that happened, the ground war could drag on for months.

The Marines expected heavy resistance. The Iraqis had fortified many of their military positions inside Kuwait, concentrating their forces near the Al Wafrah oil field along the Saudi Arabian border. They had draped a line of explosive mines across the desert. To make matters even more difficult, these Iraqi units had largely been spared the brutal air war. Because the Coalition forces were determined to minimize collateral damage and civilian casualties, bombing runs inside the occupied country were sharply restricted. Unlike the Republican Guard troops stationed in southern Iraq, a military force that had been decimated by thirty-seven days of intense bombing, these Marines were about to encounter an enemy at full strength. Central Command (CENTCOM) estimated that during the invasion of Kuwait, each Marine division would suffer approximately a thousand casualties, or between 5 and 10 percent of its total troop strength.

To support this high-stakes mission, a fleet of Coalition battleships and destroyers was positioned fewer than twenty miles off the Kuwaiti coast. This was a risky strategic move; although the big naval guns provided crucial air cover for the ground attack of Kuwait, they were also well within range of Iraqi missiles. On the morning of the Marine invasion, the American and British ships in the Persian Gulf were put on the highest possible alert. They were told to expect hostile fire.

The first twenty-four hours of the ground war exceeded even CENTCOM's high expectations. After successfully breaching the perimeter of mines and barbed wire put down by the Iraqis, the Marine division managed to penetrate deep into central Kuwait. Unlike the Soviet T-72 tanks used by the Iraqi army, the American M1 Abrams tanks were equipped with GPS units and thermal sights, allowing the Marines to engage the enemy in the pitch-black night. After a brigade of Marines reached the outskirts of Kuwait City, they made an abrupt turn to the east and began the task of securing the coastline. Just before dawn on February 25, ten Marine helicopters, along with an amphibious landing ship, conducted a feint attack on a military base near the Kuwaiti port of Ash Shuaybah. The attack was supported by a barrage of artillery rounds from the offshore battleships. The Coalition forces weren't interested in capturing the port; they just wanted to "neutralize" it, to make sure it didn't pose a danger to the offshore convoy.

That same morning, while Ash Shuaybah was being attacked, Lieutenant Commander Michael Riley was monitoring the radar screens onboard the HMS *Gloucester,* a British destroyer stationed about fifteen miles from the port. The ship was responsible for protecting the Allied fleet, which meant that Riley had to monitor all of the airspace surrounding the naval convoy. Since the start of the air war, the radar crews had maintained an exhausting schedule. They were on duty for six hours, then they had six hours to sleep and eat, and after that brief respite, they headed back to the claustrophobic radar room. By the time the ground invasion began, the men were showing signs of fatigue. They had bloodshot eyes and needed constant infusions of caffeine.

Riley had been on duty since midnight. At 5:01 in the morning, just as the Allied ships began shelling Ash Shuaybah, he noticed a radar blip off the Kuwaiti coast. A quick calculation of its trajectory had it heading straight for the convoy. Although Riley had been staring at similar-looking blips all night long, there was something about this radar trace that immediately made him suspicious. He couldn't explain why, but the blinking green dot on the screen filled him with fear; his pulse started to race and his hands became clammy. He continued to observe the incoming blip for another forty seconds as it slowly honed in on the USS *Missouri,* an American battleship. With each sweep of the radar, the blip grew closer. It was approaching the American ship at more than 550 miles per hour. If Riley was going to shoot down the target—if he was going to act on his fear—then he needed to respond right away. If that blip was a missile and Riley didn't move immediately, it would be too late. Hundreds of sailors would die. The USS *Missouri* would be sunk. And Riley would have stood by and watched it happen.

But Riley had a problem. The radar blip was located in airspace that was frequently traveled by American A-6 fighter jets, which the U.S. Navy was using to deliver laser-guided bombs to support the Marine ground invasion. After completing their sorties, the planes flew down the Kuwait coast, turned east toward the convoy, and landed on their aircraft carriers. Over the last few weeks, Riley had watched dozens of A-6s fly a route nearly identical to the one being followed by this unidentified radar blip. The blip was also traveling at the same speed as the fighter jets and had a similar surface area. It looked exactly like an A-6 on the radar screen.

To make matters even more complicated, the A-6 pilots had gotten into the bad habit of turning off their electronic identification on their return flights. This identification system allowed Coalition forces to recognize their own, but it also made the planes more vulnerable to Iraqi antiaircraft missiles. Not surprisingly, the pilots opted for the cloak of silence over Iraqi-controlled airspace. As a result, the radar crew onboard the HMS *Gloucester* had no way of contacting this radar blip.

There was one last way for radar crews to distinguish between an incoming missile and a friendly aircraft: they could determine the altitude of the blip. The A-6 generally flew at around three thousand feet, while a Silkworm missile flew at one thousand feet. However, the type of radar that Riley was using didn't provide him with any altitude information. If he wanted to know the height of a specific object, he had to use a specialized radar system known as the 909, which conducted sweeps in horizontal bands. Unfortunately, the 909 radar operator had entered an incorrect tracking number shortly after the blip appeared, which meant that Riley had no way of knowing the altitude of the flying object. Although he'd now been staring at the radar blip for almost a minute, its identity remained a befuddling mystery.

The target was moving fast. The time for deliberation was over. Riley issued the order to fire; two Sea Dart surface-to-air missiles were launched into the sky. Seconds passed. Riley nervously stared at the radar screen, watching his missiles race toward the object at speeds approaching Mach 1. The blinking green blips appeared to be drawn to the target, like iron filings to a magnet. Riley waited for the interception.

The explosion echoed over the ocean. All of the blips immediately disappeared from the radar screen. Whatever had been flying toward the USS *Missouri* helplessly fell into the sea, just seven hundred yards in front of the American battleship. A few moments later, the captain of the HMS *Gloucester* entered the radar room. "Whose bird is it?" he asked Riley, wanting to know who was responsible for destroying the still unidentified target. "It was ours, sir," Riley responded. The captain asked Riley how he could be sure he'd fired at an Iraqi missile and not at an American fighter jet. Riley said he just knew.

THE NEXT FOUR HOURS were the longest ones of Riley's life. If he had shot down an A-6, then he had killed two innocent pilots. His career was over. He might even be court-martialed. Riley immediately went back to review the radar tapes, looking for any scrap of evidence suggesting that the blip really was an Iraqi missile. But even when he had the luxury of time and analysis, Riley still couldn't definitively identify the target; the tapes were completely ambiguous. The mood on the HMS *Gloucester* quickly grew somber. Investigative teams were sent out to view the wreckage still floating on the ocean surface. An immediate inventory of all Coalition planes in the area was conducted.

The captain of the HMS *Gloucester* heard the news first. He walked over to Riley's bunk, where Riley was trying, in vain, to get some sleep. The results of the investigation were in: the radar blip was a Silkworm missile, not an American fighter jet. Riley had single-handedly saved a battleship.

Of course, it's possible that Riley had just gotten lucky. After the war was over, British naval officers carefully analyzed the sequence of events preceding Riley's decision to fire the Sea Dart missiles. They concluded that based on the radar tapes, it was impossible to distinguish between the Silkworm and a friendly A-6. Although Riley had made the correct decision, he could have just as easily been shooting down an American fighter jet. His risky gamble had paid off, but it had still been a gamble.

That, at least, was the official version of events until the summer of 1993, when Gary Klein started to investigate the Silkworm affair. A cognitive psychologist who consults for the Marine Corps, Klein was informed that nobody could explain how the radar blip had been identified as a hostile missile. Even Riley didn't know why he'd considered that early-morning blip so dangerous. He assumed, like everybody else, that he'd just gotten lucky.

Klein was intrigued. He had spent the last few decades studying decision-making in high-pressure situations, and he knew that intuition could often be astonishingly insightful, even if the origin of those insights was obscure. He was determined to find the source of Riley's fear, to figure out why this particular blip had felt so scary. So he went back to the radar tapes.

He soon realized that Riley had gotten used to seeing a very consistent blip pattern when the A-6s returned from their bombing sorties. Because Riley's naval radar could pick up signals only over water — after a signal went "wet feet" — he was accustomed to seeing the fighter jets right as they flew off the Kuwaiti coast. The planes typically became visible after a single radar sweep.

Klein analyzed the radar tapes from the predawn missile attack. He replayed those fateful forty seconds over and over again, searching for any differences between Riley's experience of the A-6s returning from their sorties and his experience of the Silkworm blip.

That's when Klein suddenly saw the discrepancy. It was subtle, but crystal clear. He could finally explain Riley's intuitive insight.

The secret was the timing. Unlike the A-6, the Silkworm didn't appear off the coast right away. Because it traveled at such a low altitude, nearly two thousand feet below an A-6's, the signal of the missile was initially masked by ground interference. As a result, it wasn't visible until the *third* radar sweep, which was eight seconds after an A-6 would have appeared. Riley was unconsciously evaluating the altitude of the blip, even if he didn't know he was doing it.

This is why Riley got the chills when he stared at the Iraqi missile on his radar screen. There was something strange about this radar blip. It didn't feel like an A-6. Although Riley couldn't explain why he felt so scared, he knew that something scary was happening. This blip needed to be shot down.

# 1

The question still remains: how did Riley's emotions manage to distinguish between these two seemingly identical radar blips? What was happening inside his brain when he first saw the Silkworm missile, three sweeps off the Kuwaiti coast? Where did his fear come from? The answer lies in a single molecule, called dopamine, that brain cells use to communicate with one another. When Riley stared at the radar screen, it was most likely his dopamine neurons that told him he was looking at a missile and not an A-6 fighter jet.

The importance of dopamine was discovered by accident. In 1954, James Olds and Peter Milner, two neuroscientists at McGill University, decided to implant an electrode deep into the center of a rat's brain. The precise placement of the electrode was largely happenstance; at the time, the geography of the mind remained a mystery. But Olds and Milner got lucky. They inserted the needle right next to the nucleus accumbens (NAcc), a part of the brain that generates pleasurable feelings. Whenever you eat a piece of chocolate cake, or listen to a favorite pop song, or watch your favorite team win the World Series, it is your NAcc that helps you feel so happy.

But Olds and Milner quickly discovered that too much pleasure can be fatal. They placed the electrodes in several rodents' brains and then ran a small current into each wire, making the NAccs continually excited. The scientists noticed that the rodents lost interest in everything. They stopped eating and drinking. All courtship behavior ceased. The rats would just huddle in the corners of their cages, transfixed by their bliss. Within days, all of the animals had perished. They died of thirst.

It took several decades of painstaking research, but neuroscientists eventually discovered that the rats had been suffering from an excess of dopamine. The stimulation of the NAcc triggered a massive release of the neurotransmitter, which overwhelmed the rodents with ecstasy. In humans, addictive drugs work the same way: a crack addict who has just gotten a fix is no different than a rat in an electrical rapture. The brains of both creatures have been blinded by pleasure. This, then, became the dopaminergic cliché; it was the chemical explanation for sex, drugs, and rock and roll.

But happiness isn't the only feeling that dopamine produces. Scientists now know that this neurotransmitter helps to regulate *all* of our emotions, from the first stirrings of love to the most visceral forms of disgust. It is the common neural currency of the mind, the molecule that helps us decide among alternatives. By looking at how dopamine works inside the brain, we can see why feelings are capable of providing deep insights. While Plato disparaged emotions as irrational and untrustworthy—the wild horses of the soul—they actually reflect an enormous amount of invisible analysis.

Much of our understanding of the dopamine system comes from the pioneering research of Wolfram Schultz, a neuroscientist at Cambridge University. He likes to compare dopamine neurons (those neurons that use dopamine to communicate) to the photoreceptors on the retina, which detect the rays of light entering the eye. Just as the process of sight starts with the retina, so the process of decision-making begins with the fluctuations of dopamine.

As a medical student in the early 1970s, Schultz grew interested in the neurotransmitter because of its role in triggering the paralyzing symptoms of Parkinson's disease. He recorded from cells in the monkey brain, hoping to find which cells were involved in controlling the body's movements. But he couldn't find anything. "It was a classic case of experimental failure," he says. "I was a very frustrated scientist." But after years of searching, Schultz noticed something odd about these dopamine neurons: they began to fire just before the monkey was given a reward, such as a pellet of food or a bit of banana. (The rewards were used to get the monkeys to move.) "At first I thought it was unlikely that an individual cell could represent anything so complicated as food," Schultz says. "It just seemed like too much information for one neuron."

After hundreds of experimental trials, Schultz began to believe his own data; he realized he had accidentally found the reward mechanism at work in the primate brain. In the mid-1980s, after publishing a series of landmark papers, Schultz set out to decipher this reward circuitry. How exactly did a single cell manage to represent a reward? And why did it fire *before* a reward was given?

The Schultz experiments followed a simple protocol: he sounded a loud tone, waited for a few seconds, and then squirted some drops of apple juice into the mouth of a monkey. While the experiment was unfolding, Schultz was probing the monkey brain with a needle that monitored the electrical activity inside individual cells. At first, the dopamine neurons fired only when the juice was delivered. The cells were responding to the actual reward. However, once the animal learned that the tone preceded the arrival of juice—this required only a few trials—the same neurons began firing at the sound of the tone instead of at the sweet reward. Schultz called these cells "prediction neurons," since they were more concerned with *predicting* rewards than actually receiving them. (This process can be indefinitely extended: the dopamine neurons can be made to respond to a light that precedes the tone that precedes the juice, and so on.) Once this simple pattern was learned, the monkey's dopamine neurons became exquisitely sensitive to variations on it. If the cellular predictions proved correct, and the reward arrived right on time, then the primate experienced a brief surge of dopamine, the pleasure of being right. However, if the pattern was violated—if the tone was played but the juice never arrived—then the monkey's dopamine neurons decreased their firing rate. This is known as the prediction-error signal. The monkey felt upset because its predictions of juice were wrong.

What's interesting about this system is that it's all about *expectation*. Dopamine neurons constantly generate patterns based on experience: if this, then that. They learn that the tone predicts the juice, or that the light predicts the tone that predicts the juice. The cacophony of reality is distilled into models of correlation that allow the brain to anticipate what will happen next. As a result, the monkeys quickly learn when to expect their sweet reward.

After refining this set of cellular forecasts, the brain compares these predictions to what actually happens. Once the monkey is taught to expect juice after a certain sequence of events, its dopamine cells carefully monitor the situation. If everything goes according to plan, its dopamine neurons secrete a little burst of enjoyment. The monkey is happy. But if these expectations aren't met—if the monkey doesn't get the promised juice—the dopamine cells go on strike. They instantly send out a signal announcing their mistake and stop releasing dopamine.

The brain is designed to amplify the shock of these mistaken predictions. Whenever it experiences something unexpected—like a radar blip that doesn't fit the usual pattern, or a drop of juice that doesn't arrive—the cortex immediately takes notice.

Within milliseconds, the activity of the brain cells has been inflated into a powerful emotion. Nothing focuses the mind like surprise.

This fast cellular process begins in a tiny area in the center of the brain that is dense with dopamine neurons. Neuroscientists have known for several years that this region, the anterior cingulate cortex (ACC), is involved in the detection of errors. Whenever the dopamine neurons make a mistaken prediction—when they expect juice but don't get it—the brain generates a unique electrical signal, known as error-related negativity. The signal emanates from the ACC, so many neuroscientists refer to this area as the "oh, shit!" circuit.

The importance of the ACC is revealed by the layout of the brain. Like the orbitofrontal cortex, the ACC helps control the conversation between what we know and what we feel. It sits at the crucial intersection between these two different ways of thinking. On the one hand, the ACC is closely connected to the thalamus, a brain area that helps direct conscious attention. This means that if the ACC is startled by some stimulus—like the bang of a gunshot it didn't expect—it can immediately focus on the relevant sensation. It forces the individual to notice the unexpected event.

While the ACC is alerting the consciousness, it's also sending signals to the hypothalamus, which regulates crucial aspects of bodily function. When the ACC is worried about some anomaly—for instance, an errant blip on a radar screen—that worry is immediately translated into a somatic signal as the muscles prepare for action. Within seconds, heart rate increases, and adrenaline pours into the bloodstream. These fleshly feelings compel us to respond to the situation *right away.* A racing pulse and sweaty palms are the brain's way of saying that there's no time to waste. This prediction error is urgent.

But the ACC doesn't just monitor erroneous predictions. It

also helps remember what the dopamine cells have just learned, so that expectations can be quickly adjusted in light of new events. It internalizes the lessons of real life, making sure that neural patterns are completely up to date. If it was predicted that juice would arrive after the tone, but the juice never arrived, then the ACC makes sure that future predictions are revised. The short-term feeling is translated into a long-term lesson. Even if the monkey is unaware of what, exactly, the ACC has memorized, the next time it's waiting for a squirt of juice, its brain cells are prepared. They know exactly when the reward will arrive.

This is an essential aspect of decision-making. If we can't incorporate the lessons of the past into our future decisions, then we're destined to endlessly repeat our mistakes. When the ACC is surgically removed from the monkey brain, the behavior of the primate becomes erratic and ineffective. The monkeys can no longer predict rewards or make sense of their surroundings. Researchers at Oxford performed an elegant experiment that made this deficit clear. A monkey clutched a joystick that moved in two different directions: it could be either lifted or turned. At any given moment, only one of the movements would trigger a reward (a pellet of food). To make things more interesting, the scientists switched the direction that would be rewarded every twenty-five trials. If the monkey had previously gotten in the habit of lifting the joystick in order to get a food pellet, it now had to shift its strategy.

So what did the monkeys do? Animals with intact ACCs had no problem with the task. As soon as they stopped receiving rewards for lifting the joystick, they started turning it in the other direction. The problem was soon solved, and the monkeys continued to receive their pellets of food. However, monkeys that were missing their ACCs demonstrated a telling defect. When they stopped being rewarded for moving the joystick in a certain direction, they were still able (most of the time) to change direction, just like the normal monkeys. However, they were unable to *persist* in this successful strategy and soon went back to moving the joystick in the direction that garnered no reward. They never learned how to consistently find the food, to turn a mistake into an enduring lesson. Because these monkeys couldn't update their cellular predictions, they ended up hopelessly confused by the simple experiment.

People with a genetic mutation that reduces the number of dopamine receptors in the ACC suffer from a similar problem; just like the monkeys, they are less likely to learn from negative reinforcement. This seemingly minor deficit has powerful consequences. For example, studies have found that people carrying this mutation are significantly more likely to become addicted to drugs and alcohol. Because they have difficulty learning from their mistakes, they make the same mistakes over and over. They can't adjust their behavior even when it proves self-destructive.

The ACC has one last crucial feature, which further explains its importance: it is densely populated with a very rare type of cell known as a spindle neuron. Unlike the rest of our brain cells, which are generally short and bushy, these brain cells are long and slender. They are found only in humans and great apes, which suggests that their evolution was intertwined with higher cognition. Humans have about forty times more spindle cells than any other primate.

The strange form of spindle cells reveals their unique function: their antenna-like bodies are able to convey emotions across the entire brain. After the ACC receives input from a dopamine neuron, spindle cells use their cellular velocity—they transmit electrical signals faster than any other neuron—to make sure that the rest of the cortex is instantly saturated in that specific feeling. The consequence of this is that the minor fluctuations of a single type of neurotransmitter play a huge role in guiding our actions, telling us how we should feel about what we see. "You're probably 99.9 percent unaware of dopamine release," says Read Montague, a professor of neuroscience at Baylor University. "But you're probably 99.9 percent driven by the information and emotions it conveys to other parts of the brain."

WE CAN NOW begin to understand the surprising wisdom of our emotions. The activity of our dopamine neurons demonstrates that feelings aren't simply reflections of hard-wired animal instincts. Those wild horses aren't acting on a whim. Instead, human emotions are rooted in the predictions of highly flexible brain cells, which are constantly adjusting their connections to reflect reality. Every time you make a mistake or encounter something new, your brain cells are busy changing themselves. Our emotions are deeply empirical.

Look, for example, at Schultz's experiment. When Schultz studied those juice-craving monkeys, he discovered that it took only a few experimental trials before the monkeys' neurons knew exactly when to expect their rewards. The neurons did this by continually incorporating the new information, turning a negative feeling into a teachable moment. If the juice didn't arrive, then the dopamine cells adjusted their expectations. Fool me once, shame on you. Fool me twice, shame on my dopamine neurons.

The same process is constantly at work in the human mind. Motion sickness is largely the result of a dopamine prediction error: there is a conflict between the type of motion being experienced—for instance, the unfamiliar pitch of a boat—and the type of motion *expected* (solid, unmoving ground). The result in this case is nausea and vomiting. But it doesn't take long before the dopamine neurons start to revise their models of motion; this is why seasickness is usually temporary. After a few horrible hours, the dopamine neurons fix their predictions and learn to expect the gentle rocking of the high seas.

When the dopamine system breaks down completely—when neurons are unable to revise their expectations in light of reality—mental illness can result. The roots of schizophrenia remain shrouded in mystery, but one of its causes seems to be an excess of certain types of dopamine receptors. This makes the dopamine system hyperactive and disregulated, which means that the neurons of a schizophrenic are unable to make cogent predictions or correlate their firing with outside events. (Most antipsychotic medications work by reducing the activity of dopamine neurons.) Because schizophrenics cannot detect the patterns that actually exist, they start hallucinating false patterns. This is why schizophrenics become paranoid and experience completely unpredictable shifts in mood. Their emotions have been uncoupled from the events of the real world.

The crippling symptoms of schizophrenia serve to highlight the necessity and precision of dopamine neurons. When these neurons are working properly, they are a crucial source of wisdom. The emotional brain effortlessly figures out what's going on and how to exploit the situation for maximum gain. Every time you experience a feeling of joy or disappointment, fear or happiness, your neurons are busy rewiring themselves, constructing a theory of what sensory cues preceded the emotions. The lesson is then committed to memory, so the next time you make a decision, your brain cells are ready. They have learned how to predict what will happen next.

2

Backgammon is the oldest board game in the world. It was first played in ancient Mesopotamia, starting around 3000 B.C. The game was a popular diversion in ancient Rome, celebrated by the Persians, and banned by King Louis IX of France for encouraging illicit gambling. In the seventeenth century, Elizabethan courtiers codified the rules of backgammon, and the game has changed little since.

The same can't be said about the *players* of the game. One of the best backgammon players in the world is now a software program. In the early 1990s, Gerald Tesauro, a computer programmer at IBM, began developing a new kind of artificial intelligence (AI). At the time, most AI programs relied on the brute computational power of microchips. This was the approach used by Deep Blue, the powerful set of IBM mainframes that managed to defeat chess grand master Garry Kasparov in 1997. Deep Blue was capable of analyzing more than two hundred million possible chess moves per second, allowing it to consistently select the optimal chess strategy. (Kasparov's brain, on the other hand, evaluated only about five moves per second.) But all this strategic firepower consumed a lot of energy: while playing chess, Deep Blue was a fire hazard and required specialized heat-dissipating equipment so that it didn't burst into flames. Kasparov, meanwhile, barely broke a sweat. That's because the human brain is a model of efficiency: even when it's deep in thought, the cortex consumes less energy than a light bulb.

While the popular press was celebrating Deep Blue's stunning achievement—a machine had outwitted the greatest chess player in the world!—Tesauro was puzzled by its limitations. Here was a machine capable of thinking millions of times faster than its human opponent, and yet it had barely won the match. Tesauro realized that the problem with all conventional AI programs, even brilliant ones like Deep Blue's, was their *rigidity*. Most of Deep Blue's intelligence was derived from other chess grand masters, whose wisdom was painstakingly programmed into the machine. (IBM programmers also studied Kasparov's previous chess matches and engineered the software to exploit his recurring strategic mistakes.) But the machine itself was incapable of learning. Instead, it made decisions by predicting the probable outcomes of several million different chess moves. The move with the highest predicted "value" was what the computer ended up executing. For Deep Blue, the game of chess was just an endless series of math problems.

Of course, this sort of artificial intelligence isn't an accurate model of human cognition. Kasparov managed to compete on the same level as Deep Blue even though his mind had far less computational power. Tesauro's surprising insight was that Kasparov's neurons were effective because they had trained themselves. They had been refined by decades of experience to detect subtle spatial patterns on the chessboard. Unlike Deep Blue, which analyzed *every* possible move, Kasparov was able to instantly winnow his options and focus his mental energies on evaluating only the most useful strategic alternatives.

Tesauro set out to create an AI program that acted like Garry Kasparov. He chose backgammon as his paradigm and named the program TD-Gammon. (The *TD* stands for *temporal difference*.) Deep Blue had been preprogrammed with chess acumen, but Tesauro's software began with absolutely zero knowledge. At first, its backgammon moves were entirely random. It lost every match and made stupid mistakes. But the computer didn't remain a novice for long; TD-Gammon was designed to learn from its own experience. Day and night, the software played backgammon against itself, patiently learning which moves were most effective. After a few hundred thousand games of backgammon, TD-Gammon was able to defeat the best human players in the world.

How did the machine turn itself into an expert? Although the mathematical details of Tesauro's software are numbingly complex, the basic approach is simple.* TD-Gammon generates a set of predictions about how the backgammon game will unfold. Unlike Deep Blue, the computer program doesn't investigate every possible permutation. Instead, it acts like Garry Kasparov and generates its predictions from its previous experiences. The software compares these predictions to what actually happens during the backgammon game. The ensuing discrepancies provide the substance of its education, and the software strives to continually decrease this "error signal." As a result, its predictions constantly increase in accuracy, which means that its strategic decisions get more and more effective and intelligent.

In recent years, the same software strategy has been used to solve all kinds of difficult problems, from programming banks of elevators in skyscrapers to determining the schedules of flights. "Anytime you've got a problem with a seemingly infinite number of possibilities"—the elevators and planes can be arranged in any number of sequences—"these sorts of learning programs can be a crucial guide," says Read Montague. The essential distinction between these reinforcement-learning programs and traditional approaches is that these new programs find the optimal solutions by themselves. Nobody tells the computer how to organize the elevators. Instead, it methodically learns by running trials and focusing on its errors until, after a certain number of

---

*The TD-learning model used by Tesauro was based on the pioneering work of computer scientists Rich Sutton and Andrew Barto. In the early 1980s, when they were grad students at the University of Massachusetts, Amherst, Sutton and Barto tried to develop a model of artificial intelligence that could learn simple rules and behaviors and apply them to achieve a goal. It was an audacious idea; their academic advisers tried to dissuade them from even trying, but the young scientists were stubborn. "It had always been this kind of untouchable goal in computer science," Sutton says. "Marvin Minsky had done his thesis on reinforcement learning and basically given up. He said it was impossible and left the field. Luckily for us, it wasn't impossible. We knew even simple animals could learn like this—nobody teaches a bird how to find a worm—we just didn't know how."

trials, the elevators are running as efficiently as possible. The seemingly inevitable mistakes have disappeared.

This programming method closely mirrors the activity of dopamine neurons. The brain's cells also measure the mismatch between expectation and outcome. They use their inevitable errors to improve performance; failure is eventually turned into success. Take, for example, an experiment known as the Iowa Gambling Task designed by the neuroscientists Antonio Damasio and Antoine Bechara. The game went as follows: a subject — "the player" — was given four decks of cards, two black and two red, and $2,000 of play money. Each card told the player whether he'd won or lost money. The subject was instructed to turn over a card from one of the four decks and to make as much money as possible.

But the cards weren't distributed at random. The scientists had rigged the game. Two of the decks were full of high-risk cards. These decks had bigger payouts ($100), but also contained extravagant punishments ($1,250). The other two decks, by comparison, were staid and conservative. Although they had smaller payouts ($50), they rarely punished the player. If the gambler drew only from those two decks, he would come out way ahead.

At first, the card-selection process was entirely haphazard. There was no reason to favor any specific deck, and so most people sampled from each pile, searching for the most lucrative cards. On average, people had to turn over about fifty cards before they began to draw solely from the profitable decks. It took about eighty cards before the average experimental subject could explain *why* he or she favored those decks. Logic is slow.

But Damasio wasn't interested in logic; he was interested in emotion. While the gamblers in the experiment were playing the card game, they were hooked up to a machine that measured the electrical conductance of their skin. In general, higher levels of conductance signal nervousness and anxiety. What the scientists found was that after a player had drawn only ten cards, his hand got "nervous" when it reached for the negative decks. Although the subject still had little inkling of which card piles were the most lucrative, his emotions had developed an accurate sense of fear. The emotions knew which decks were dangerous. The subject's feelings figured out the game first.

Neurologically impaired patients who were unable to experience any emotions at all— usually because of damaged orbitofrontal cortices—proved incapable of selecting the right cards. While most people made substantial amounts of money during the experiment, these purely rational people often went bankrupt and had to take out "loans" from the experimenter. Because these patients were unable to associate the bad decks with negative feelings—their hands never developed the symptoms of nervousness—they continued to draw equally from all four decks. When the mind is denied the emotional sting of losing, it never figures out how to win.

How do emotions become so accurate? How do they identify the lucrative decks so quickly? The answer returns us to dopamine, the molecular source of our feelings. By playing the Iowa Gambling Task with a person undergoing brain surgery for epilepsy—the patient was given local anesthesia but remained awake during the surgery—scientists at the University of Iowa and Caltech were able to watch the learning process unfold in real time. The scientists discovered that human brain cells are programmed just like TD-Gammon: they generate predictions about what will happen and then measure the difference between their expectations and the actual results. In the Iowa Gambling Task experiment, if a cellular prediction proved false—for example, if the player chose the bad deck—then the dopamine neurons immediately stopped firing. The player experienced a negative emotion and learned not to draw from that deck again.

(Disappointment is educational.) However, if the prediction was accurate—if he got rewarded for choosing a lucrative card—then the player felt the pleasure of being correct, and that particular connection was reinforced. As a result, his neurons quickly learned how to make money. They had found the secret to winning the gambling game before the player could understand and explain the solution.

This is a crucial cognitive talent. Dopamine neurons automatically detect the subtle patterns that we would otherwise fail to notice; they assimilate all the data that we can't consciously comprehend. And then, once they come up with a set of refined predictions about how the world works, they translate these predictions into emotions. Let's say, for example, that you're given lots of information about how twenty different stocks have performed over a period of time. (The various share prices are displayed on a ticker tape at the bottom of a television screen, just as they appear on CNBC.) You'll soon discover that you have difficulty remembering all the financial data. If somebody asks you which stocks performed the best, you'll probably be unable to give a good answer. You can't process all the information. However, if you're asked which stocks trigger the best *feelings*—your emotional brain is now being quizzed—you'll suddenly be able to identify the best stocks. According to Tilmann Betsch, the psychologist who performed this clever little experiment, your emotions will "reveal a remarkable degree of sensitivity" to the actual performance of all of the different securities. The investments that rose in value will be associated with the most positive emotions, while the shares that went down in value will trigger a vague sense of unease. These wise yet inexplicable feelings are an essential part of the decision-making process. Even when we think we know nothing, our brains know something. That's what our feelings are trying to tell us.

3

This doesn't mean that people can coast on these cellular emotions. Dopamine neurons need to be continually trained and retrained, or else their predictive accuracy declines. Trusting one's emotions requires constant vigilance; intelligent intuition is the result of deliberate practice. What Cervantes said about proverbs—"They are short sentences drawn from long experience"—also applies to brain cells, but only if we use them properly.

Consider Bill Robertie. He's one of the only people in the world who's a world-class expert in three different games. (Imagine if Bo Jackson had played in the NBA in addition to the NFL and baseball's major leagues . . .) Robertie is a chess master and a former winner of the U.S. speed chess championship. He's a widely respected poker expert and best-selling author of several books on Texas hold'em. However, Robertie is best known for his backgammon skills. He has won the World Championship of Backgammon twice (a feat accomplished by only one other person), and is regularly ranked among the top twenty players in the world. In the early 1990s, when Gerald Tesauro was looking for a backgammon expert to compete against TD-Gammon, he chose Robertie. "He wanted the computer to learn from the best," Robertie says. "And I was the best."

Robertie is now in his early sixties, with a shock of graying hair, lidded eyes, and a pair of thick spectacles. He managed to turn a childhood obsession with chess into a lucrative career. When Robertie talks about games, he still speaks with the boyish enthusiasm of someone who can't quite believe that he gets to play for a living. "The first time I competed against TD-Gammon I was incredibly impressed," Robertie says. "It represented a big improvement over any other computer program I'd ever encountered. But I knew that I was still a better player. The next year, however, was a different story. The computer was now a really formidable opponent. It had learned how to play from playing me."

The software program became a backgammon expert by studying its prediction errors. After making a few million mistakes, TD-Gammon was able to join the shortlist of computers, like Deep Blue, that are able to compete with the best human opponents. However, all of these brilliant machines come with a strict limitation: they can each master only one game. TD-Gammon can't play chess, and Deep Blue can't play backgammon. No computer has been able to master poker.

So how did Robertie get so good at such different games? At first glance, chess, backgammon, and poker seem to rely on very different cognitive skills. That's why most backgammon champions tend to play nothing but backgammon; most chess masters don't bother with card games; and most poker players couldn't tell a Latvian Gambit from a French Defense. And yet, Robertie manages to excel in all three domains. According to Robertie, his success has a simple explanation: "I know how to practice," he says. "I know how to make myself better."

In the early 1970s, when Robertie was still just a chess prodigy—he made a living by winning speed chess tournaments—he stumbled upon backgammon. "Right away, I fell in love with the game," he says. "Plus, there was a lot more money in backgammon than speed chess." Robertie bought a book on backgammon strategy, memorized a few opening moves, and then started to play. And play. And play. "You've got to get obsessed," he says. "You've got to reach the point where you're having dreams about the game."

After a few years of intense practice, Robertie had turned himself into one of the best backgammon players in the world. "I knew I was getting good when I could just glance at a board and know what I should do," Robertie says. "The game started to become very much a matter of aesthetics. My decisions increasingly depended on the look of things, so that I could contemplate a move and then see right away if it made my position look better or worse. You know how an art critic can look at a painting and just know if it's a good painting? I was the same way, only my painting was the backgammon board."

But Robertie didn't become a world champion just by playing a lot of backgammon. "It's not the quantity of practice, it's the *quality,*" he says. According to Robertie, the most effective way to get better is to focus on your mistakes. In other words, you need to consciously consider the errors being internalized by your dopamine neurons. After Robertie plays a chess match, or a poker hand, or a backgammon game, he painstakingly reviews what happened. Every decision is critiqued and analyzed. Should he have sent out his queen sooner? Tried to bluff with a pair of sevens? What if he had consolidated his backgammon blots? Even when Robertie wins—and he almost always wins—he insists on searching for his errors, dissecting those decisions that could have been a little bit better. He knows that self-criticism is the secret to self-improvement; negative feedback is the best kind. "That's one of the things I learned from TD-Gammon," Robertie says. "Here was a computer that did nothing but measure what it got wrong. That's all it did. And it was as good as me."

The physicist Niels Bohr once defined an expert as "a person who has made all the mistakes that can be made in a very narrow field." From the perspective of the brain, Bohr was absolutely right. Expertise is simply the wisdom that emerges from cellular error. Mistakes aren't things to be discouraged. On the contrary, they should be cultivated and carefully investigated.

Carol Dweck, a psychologist at Stanford, has spent decades demonstrating that one of the crucial ingredients of successful education is the ability to learn from mistakes. The same strategy that Robertie uses to excel at games is also an essential pedagogic tool. Unfortunately, children are often taught the exact opposite. Instead of praising kids for trying hard, teachers typically praise them for their innate intelligence (being smart). Dweck has shown that this type of encouragement actually backfires, since it leads students to see mistakes as signs of stupidity and not as the building blocks of knowledge. The regrettable outcome is that kids never learn how to learn.

Dweck's most famous study was conducted in twelve different New York City schools and involved more than four hundred fifth-graders. One at a time, the kids were removed from class and given a relatively easy test consisting of nonverbal puzzles. After the child finished the test, the researchers told the student his or her score and provided a single sentence of praise. Half of the kids were praised for their *intelligence*. "You must be smart at this," the researcher said. The other students were praised for their *effort*: "You must have worked really hard."

The students were then allowed to choose between two different subsequent tests. The first choice was described as a more difficult set of puzzles, but the kids were told that they'd learn a lot from attempting it. The other option was an easy test, similar to the test they'd just taken.

When Dweck was designing the experiment, she'd expected the different forms of praise to have a rather modest effect. After all, it was just one sentence. But it soon became clear that the type of compliment given to the fifth-graders dramatically influenced their choice of tests. Of the group of kids that had been praised for their efforts, 90 percent chose the harder set of puzzles. However, of the kids that were praised for their intelligence, most went for the easier test. "When we praise children for their intelligence," Dweck wrote, "we tell them that this is the name of the game: Look smart, don't risk making mistakes."

Dweck's next set of experiments showed how this fear of failure actually inhibited learning. She gave the same fifth-graders yet another test. This test was designed to be extremely difficult—it was originally written for eighth-graders—but Dweck wanted to see how the kids would respond to the challenge. The students who had been praised for their efforts in the initial test worked hard at figuring out the puzzles. "They got very involved," Dweck says. "Many of them remarked, unprovoked, 'This is my favorite test.'" Kids that had initially been praised for their smarts, on the other hand, were easily discouraged. Their inevitable mistakes were seen as signs of failure: perhaps they really weren't smart after all. After taking this difficult test, the two groups of students had to choose between looking at the exams of kids who did worse than them and looking at the exams of those who did better. Students praised for their intelligence almost always chose to bolster their self-esteem by comparing themselves with students who had performed worse on the test. In contrast, kids praised for their hard work were more interested in the higher-scoring exams. They wanted to understand their mistakes, to learn from their errors, to figure out how to do better.

The final round of tests was the same difficulty level as the initial test. Nevertheless, students who'd been praised for their efforts exhibited significant improvement, raising their average score by 30 percent. Because these kids were willing to challenge themselves, even if it meant failing at first, they ended up performing at a much higher level. This result was even more impressive when compared with students who'd been randomly assigned to the "smart" group; they saw their scores drop by an average of nearly 20 percent. The experience of failure had been so discouraging for the "smart" kids that they actually regressed.

The problem with praising kids for their innate intelligence —the "smart" compliment—is that it misrepresents the neural reality of education. It encourages kids to avoid the most useful kind of learning activities, which is learning from mistakes.

Unless you experience the unpleasant symptoms of being wrong, your brain will never revise its models. Before your neurons can succeed, they must repeatedly fail. There are no shortcuts for this painstaking process.

This insight doesn't apply only to fifth-graders solving puzzles; it applies to everyone. Over time, the brain's flexible cells become the source of expertise. Although we tend to think of experts as being weighed down by information, their intelligence dependent on a vast amount of explicit knowledge, experts are actually profoundly intuitive. When an expert evaluates a situation, he doesn't systematically compare all the available options or consciously analyze the relevant information. He doesn't rely on elaborate spreadsheets or long lists of pros and cons. Instead, the expert naturally depends on the emotions generated by his dopamine neurons. His prediction errors have been translated into useful knowledge, which allows him to tap into a set of accurate feelings he can't begin to explain.

The best experts embrace this intuitive style of thinking. Bill Robertie makes difficult backgammon decisions by just "looking" at the board. Thanks to his rigorous practice techniques, he's confident that his mind has already internalized the ideal moves. Garry Kasparov, the chess grand master, obsessively studied his past matches, looking for the slightest imperfection, but when it came time to play a chess game, he said he played by instinct, "by smell, by feel." After Herb Stein finishes shooting a soap opera episode, he immediately goes home and reviews the rough cut. "I watch the whole thing," Stein says, "and I just take notes. I'm looking really hard for my mistakes. I pretty much always want to find thirty mistakes, thirty things that I could have done better. If I can't find thirty, then I'm not looking hard enough." These mistakes are usually little things, so minor that nobody else would notice. But Stein knows that the only way to get it right the next time is to study what he got wrong this time.

Tom Brady spends hours watching game tape every week, critically looking at each of his passing decisions, but when he's standing in the pocket he knows that he can't hesitate before making a throw. It's not an accident that all of these experts have converged on such a similar method. They have figured out how to take advantage of their mental machinery, to steal as much wisdom as possible from their inevitable errors.

And then there's Lieutenant Commander Michael Riley. Before becoming an officer in the Royal Navy, Riley had spent years learning how to interpret the ambiguous blips on a radar screen. In the Royal Navy, the training process for such warfare specialists revolves around realistic battle simulations so that senior lieutenants like Riley can practice decision-making in its proper context. Officers are able to learn from their mistakes without having to shoot anything down.

During the Persian Gulf War, all of this training paid off. Even though Riley had never seen a Silkworm missile before, his mind had learned how to detect it. Because he had been staring at a radar screen for weeks on end, watching dozens of A-6 jets return from sorties off the Kuwaiti coast, Riley's dopamine neurons started to anticipate a consistent sequence of events. The radar pattern of the American planes had been seared into his brain. But then, in the predawn hours following the ground invasion, Riley saw a radar blip that looked slightly different. When the incoming unidentified blip appeared, it was too far out to sea, three sweeps away from the coast. As a result, a dopamine neuron somewhere in Riley's midbrain was surprised. Here was something that didn't fit the pattern, an error of expectation. The cell instantly responded to the surprising turn of events and altered its rate of firing. This electrical message was passed from neuron to neuron until it reached the ACC. Spindle cells publicized this prediction error all over the brain. Riley's years of naval training were summarized in a single flash of fear.

It was just a feeling, but Riley dared to trust it. "Fire two Sea Darts!" he yelled. The defensive missiles were launched into the sky. The battleship was saved.

SO FAR, we've been exploring the surprising intelligence of our emotions. We've seen how the fluctuations of dopamine are translated into a set of prophetic feelings. But emotions aren't perfect. They are a crucial cognitive tool, but even the most useful tools can't solve every problem. In fact, there are certain conditions that consistently short-circuit the emotional brain, causing people to make bad decisions. The best decision-makers know which situations require *less* intuitive responses, and in the next part of the book, we'll look at what those situations are.

# 3

# Fooled by a Feeling

Ann Klinestiver was working as a high school English teacher in a small town in West Virginia when she was diagnosed with Parkinson's disease. She was only fifty-two years old, but the symptoms were unmistakable. While she was standing at the front of her class trying to teach her students some Shakespeare, her hands started to shake uncontrollably. Then her legs went limp. "I lost control of my own body," she says. "I'd look at my arm, and I'd tell it what to do, but it just wouldn't listen."

Parkinson's is a disease of the dopamine system. It begins when dopamine neurons start to die in a part of the brain that controls the body's movements. Nobody knows why these cells die, but once they are gone, the loss is irrevocable. By the time the symptoms of Parkinson's appear, more than 80 percent of these neurons will be dead.

Ann's neurologist immediately put her on Requip, a drug that imitates the activity of dopamine in the brain. (It's part of a class of drugs called dopamine agonists.) While there are many different treatments for Parkinson's patients, all operate on a similar principle: increase the amount of dopamine in the brain. By making the few surviving dopamine neurons more effective at transmitting dopamine, these medicines help compensate for the massive cell death. They allow a faint electrical signal to break through the ravages of the disease. "At first, the drug was like a miracle," Ann says. "All my movement problems just disappeared." Over time, however, Ann was forced to take higher and higher doses of Requip in order to quiet her tremors. "You can feel your brain going," she says. "I became completely dependent on this drug just to get myself out of bed and put on my clothes. I needed it to live my life."

That's when Ann discovered slot machines. It was an unlikely attraction. "I'd never been interested in gambling," Ann says. "I'd always avoided casinos. My daddy was a Christian, and he raised me to believe that gambling was a sin, that it was something you were never supposed to do." But after she started taking the dopamine agonist, Ann found the slots at her local dog-racing track completely irresistible. She started gambling as soon as the track opened, at seven in the morning, and kept playing the machines until three thirty the next morning, when the security guards kicked her out. "Then I would go back home and gamble on the Internet until I could get back to the real machines," she says. "I was able to keep that up for two or three days at a time." After each of her gambling binges, Ann always swore to stay away. Sometimes, she was even able to stop gambling for a day or two. But then she'd find herself back at the racetrack, sitting in front of the slot machine, gambling away everything she had.

After a year of addictive gambling, Ann had lost more than $250,000. She had exhausted her retirement savings and emptied her pension fund. "Even when I had no money left, I still couldn't stop gambling," she says. "I was living on peanut butter, straight from the jar. I sold everything I could sell. My silverware, my clothes, my television, my car. I pawned my diamond ring. I knew I was destroying my life, but I just couldn't stop. There's no worse feeling than that."

Ann's husband eventually left her. He promised to return if she got control of her gambling habit, but Ann kept relapsing. He would find her at the track in the middle of the night, hunched in front of a slot machine, a bucket of coins in her lap and a bag of groceries on the floor. "I was a shell of a person," she says. "I stole quarters from my grandkids. I lost everything that mattered."

In 2006, Ann was finally taken off her dopamine agonist. Her movement problems came back, but the gambling compulsion immediately disappeared. "I haven't gambled in eighteen months," she says, with more than a little pride in her voice. "I still think about the slots, but the obsession isn't there. Without the drug, I don't need to play those damn machines. I'm free."

Klinestiver's sad story is disturbingly common. Medical studies suggest that as many as 13 percent of patients taking dopamine agonists develop severe gambling compulsions. People with no history of gambling suddenly become addicts. While most of these people obsess over slot machines, others get hooked on Internet poker or blackjack. They squander everything they have on odds that are stacked against them.*

Why does an excess of dopamine in a few neurons make games of chance so irresistible? The answer reveals a serious flaw in the human brain, which casinos have learned to exploit. Think how a slot machine works: You put in a coin and pull the lever. The reels start to whir. Pictures of cherries and diamonds and figure sevens fly by. Eventually, the machine settles on its verdict. Since slot machines are programmed to return only about 90

---

*Slot machines account for about 70 percent of the $48 billion a year Americans spend at casinos, which means that the average citizen spends five times more on slot machines than he or she does on movie tickets. There are now twice as many slot machines as ATMs in America.

percent of wagered money over the long term, chances are you lost money.

Now think about the slot machine from the perspective of your dopamine neurons. The purpose of these cells is to predict future events. They always want to know what occurrences — a loud tone, a flashing light, and so forth — will precede the arrival of the juice. While you are playing the slots, putting quarter after quarter into the one-armed bandit, your neurons are struggling to decipher the patterns inside the machine. They want to understand the game, to decode the logic of luck, to find the circumstances that predict a payout. So far, you're acting just like a monkey trying to predict when his squirt of juice is going to arrive.

But here's the catch: while dopamine neurons get excited by predictable rewards — they increase their firing when the juice arrives *after* the loud tone that heralded it — they get even more excited by surprising ones. According to Wolfram Schultz, such unpredictable rewards are typically three to four times more exciting, at least for dopamine neurons, than rewards that can be predicted in advance. (In other words, the best-tasting juice is the juice that was most unexpected.) The purpose of this dopamine surge is to make the brain pay attention to new, and potentially important, stimuli. Sometimes this cellular surprise can trigger negative feelings, such as fear, as happened to Lieutenant Commander Michael Riley. In the casino, however, the sudden burst of dopamine is intensely pleasurable, since it means that you've just won some money.

Most of the time, the brain will eventually get over its astonishment. It'll figure out which events predict the reward, and the dopamine neurons will stop releasing so much of the neurotransmitter. The danger of slot machines, however, is that they are inherently unpredictable. Because they use random number generators, there are no patterns or algorithms to uncover. (There is only a stupid little microchip churning out arbitrary digits.) Even though the dopamine neurons try to make sense of the rewards —they want to know when to expect some coins in return for all those squandered quarters—they keep getting surprised.

At this point, the dopamine neurons should just surrender: the slot machine is a waste of mental energy. They should stop paying attention to the surprising rewards, because the appearance of the rewards will always be surprising. But this isn't what happens. Instead of getting bored by the haphazard payouts, the dopamine neurons become obsessed. When you pull the lever and get a reward, you experience a rush of pleasurable dopamine, precisely because the reward was so unexpected, because your brain cells had no idea what was about to happen. The clanging coins and flashing lights are like a surprise squirt of juice. Because the dopamine neurons can't figure out the pattern, they can't adapt to the pattern. The result is that you are transfixed by the slot machine, riveted by the fickle nature of its payouts.

For Parkinson's patients on dopamine agonists, the surprising rewards of the casino trigger a massive release of chemical bliss. Their surviving dopamine neurons are so full of dopamine that the neurotransmitter spills over and pools in the empty spaces between cells. The brain is flooded with a feel-good chemical, making these games of chance excessively seductive. Such patients are so blinded by the pleasures of winning that they slowly lose everything. That's what happened to Ann.

The same science that revealed the importance of emotions to making decisions—Tom Brady finds the open man by listening to his feelings—is also beginning to show us the dark side of feeling too deeply. While the emotional brain is capable of astonishing wisdom, it's also vulnerable to certain innate flaws. These are the situations that cause the horses in the human mind to run wild, so that people gamble on slot machines and pick the wrong stocks and run up excessive credit card bills. When emotions get out of control—and there are certain things that reliably make this happen — the results can be just as devastating as not having any emotions at all.

# 1

In the early 1980s, the Philadelphia 76ers were one of the greatest teams in NBA history. The center of the team was Moses Malone, voted Most Valuable Player in the league. He dominated the paint, averaging twenty-five points and fifteen rebounds per game. The power forward was Julius Erving, a future Hall of Famer, who pioneered the modern style of basketball play with his elegant drives and extravagant slam dunks. In the backcourt were Andrew Toney — his accurate jump shot was a constant offensive threat — and Maurice Cheeks, one of the league leaders in assists and steals.

The 76ers entered the 1982 playoffs with the best record in the NBA. Before the first round of the postseason, a reporter asked Malone what the 76ers thought of their competition. His answer made headlines: "Four, four, four," he said, suggesting that the team would sweep all of their opponents. That had never been done before.

Malone's audacious prediction wasn't far off. During the playoffs, the 76ers' team was like a scoring machine. The offense ran through Malone in the post, but if Malone was double-teamed he simply had to swing the ball over to Erving or kick it out to Toney for a jumper. At times, the players seemed to be incapable of missing shots. On their way to the championship, the 76ers lost one game only, in the second round to Milwaukee. A slightly amended version of Malone's prediction was inscribed on the championship rings: "Fo, five, fo." It was one of the most dominant team performances in basketball history.

While the 76ers were prevailing in the postseason, the psychologists Amos Tversky and Thomas Gilovich were thinking about the imperfections of the human mind. Tversky would later recall watching the NBA games and hearing the television announcers talk about various kinds of streaks. For instance, the sportscasters alluded to the "hot hand" of Julius Erving and said that Andrew Toney was "in the zone." By the time the 76ers reached the NBA finals, the temperature of the team had become a cliché. How could they possibly lose when they were on such a roll?

But all this talk of hot hands and streaks made Tversky and Gilovich curious. Had Moses Malone really become so unstoppable? Could Andrew Toney really not miss a shot? Were the 76ers really as invincible as everyone said? So Tversky and Gilovich decided to conduct a little research experiment. Their question was simple: do players make more shots when they are hot, or do people just *imagine* that they make more shots? In other words, is the hot hand a real phenomenon?

Tversky and Gilovich began the investigation by sifting through years of 76er statistics. They looked at every single shot taken by every single player and then recorded if that shot had been preceded by a string of hits or misses. (The 76ers were one of the few NBA teams that kept track of the order in which shots were taken.) If the hot hand was a real phenomenon, then a hot player should have a higher field-goal percentage after making several previous shots. The streak should elevate his game.

So what did the scientists find? There was absolutely no evidence of the hot hand. A player's chance of making a shot was not affected by whether or not his previous shots had gone in. Each field-goal attempt was its own independent event. The short runs experienced by the 76ers were no different than the short runs that naturally emerge from any random process. Taking a jumper was like flipping a coin. The streaks were a figment of the imagination.

The 76ers were shocked by the evidence. Andrew Toney, the shooting guard, was particularly hard to convince: he was sure that he was a streaky shooter who went through distinct hot and cold periods. But the statistics told a different story. During the regular season, Toney made 46 percent of all his shots. After hitting three shots in a row—a sure sign that he was "in the zone" —Toney's field-goal percentage *dropped* to 34 percent. When Toney thought he was hot, he was actually freezing cold. And when he thought he was cold, he was just getting warmed up: after missing three shots in a row, Toney made 52 percent of his shots, which was significantly higher than his normal average.

But maybe the 76ers' team was a statistical outlier. After all, according to a survey conducted by the scientists, 91 percent of serious NBA fans believed in the hot hand. They just *knew* that players were streaky. So Tversky and Gilovich decided to analyze another basketball team: the Boston Celtics. This time, they looked at free-throw attempts too, not just field goals. Once again, they found absolutely no evidence of hot hands. Larry Bird was just like Andrew Toney: after he made several free throws in a row, his free-throw percentage actually declined. Bird got complacent and started missing shots he should have made.

Why do we believe in streaky shooters? Our dopamine neurons are to blame. Although these cells are immensely useful —they help us predict events that are actually predictable—they can also lead us astray, especially when we are confronted with randomness. Look, for example, at this elegant little experiment: A rat was put in a T-shaped maze with a few morsels of food placed on either the far right or the far left side of the enclosure. The placement of the food was random, but the dice were rigged: over the long run, the food was placed on the left side 60 percent of the time. How did the rat respond? It quickly realized that the left side was more rewarding. As a result, it always went to the left of the maze, which resulted in a 60 percent success rate. The rat didn't strive for perfection. It didn't search for a unified theory of the T-shaped maze. It just accepted the inherent uncertainty of the reward and learned to settle for the option that usually gave the best outcome.

The experiment was repeated with Yale undergraduates. Unlike the rat, the students, with their elaborate networks of dopamine neurons, stubbornly searched for the elusive pattern that determined the placement of the reward. They made predictions and then tried to learn from their prediction errors. The problem was that there was nothing to predict; the apparent randomness was real. Because the students refused to settle for a 60 percent success rate, they ended up with a 52 percent success rate. Although most of the students were convinced that they were making progress toward identifying the underlying algorithm, they were, in actuality, outsmarted by a rat.

The danger of random processes—things like slot machines and basketball shots—is that they take advantage of a defect built into the emotional brain. Dopamine neurons get such a visceral thrill from watching a hot player sink another jumper or from winning a little change from a one-armed bandit or from correctly guessing the placement of a food morsel that our brains completely misinterpret what's actually going on. We trust our feelings and perceive patterns, but the patterns don't actually exist.

Of course, it can be extremely hard to reconcile perceptions of streaks and runs with the statistical realities of an unruly world. When Apple first introduced the shuffle feature on its iPods, the shuffle was truly random; each song was equally as likely to get picked as any other. However, the randomness didn't *appear* random, since some songs were occasionally repeated, and customers concluded that the feature contained some secret patterns and preferences. As a result, Apple was forced to revise the algorithm. "We made it less random to make it feel more random," said Steve Jobs, the CEO of Apple.* Or consider Red Auerbach, the legendary Celtics coach. After being told about Tversky's statistical analysis of the hot hand, he reportedly responded with a blunt dismissal. "So he makes a study," Auerbach said. "I couldn't care less."† The coach refused to consider the possibility that the shooting streaks of the players might be a fanciful invention of his brain.

But Auerbach was wrong to disregard the study; the belief in illusory patterns seriously affects the flow of basketball games. If a team member had made several shots in a row, he was more likely to get the ball passed to him. The head coach would call a new set of plays. Most important, a player who thinks he has a hot hand has a distorted sense of his own talent, which leads him to take riskier shots, since he assumes his streak will save him. (It's the old bane of overconfidence.) Of course, the player is also more likely to miss these riskier shots. According to Tversky and Gilovich, the best shooters always think they're cold. When their

---

*This misconception is known as the gambler's fallacy. It occurs when people assume that an event is more or less likely to occur based on whether or not that event has recently occurred. As a result, people are surprised when a shuffled song repeats or when a flipped coin exhibits extended streaks of heads or tails. The most famous example of such a phenomenon occurred in a Monte Carlo casino in the summer of 1913 when a roulette wheel landed on black *twenty-six* times in a row. During that staggeringly improbable run, most gamblers bet against black, since they felt that the red must be "due." In other words, they assumed that the randomness of the roulette wheel would somehow correct the imbalance and cause the wheel to land on red. The casino ended up making millions of francs.

†Thomas Gilovich also looked at the reactions of London residents during the Blitz of 1940. While the Blitz was happening, British newspapers published maps that displayed the precise location of every German missile strike. The problem was that the strikes didn't look random, which led London residents and British military planners to conclude that the Germans could aim their missiles at specific targets. As a result, people fled those neighborhoods that seemed hardest hit and suspected that German spies lived in the areas that were mostly spared. The reality, however, was that the German military had virtually no control over where the missiles ended up. Although they aimed for central London, they were completely unable to target locations *within* London. The patterns of damage were utterly random.

feelings tell them to take the shots because they've got the hot hands, they don't listen.

THIS DEFECT IN the emotional brain has important consequences. Think about the stock market, which is a classic example of a random system. This means that the past movement of any particular stock cannot be used to predict its future movement. The inherent randomness of the market was first proposed by the economist Eugene Fama in the early 1960s. Fama looked at decades of stock-market data in order to prove that no amount of knowledge or rational analysis could help anyone figure out what would happen next. All of the esoteric tools used by investors to make sense of the market were pure nonsense. Wall Street was like a slot machine.

The danger of the stock market, however, is that sometimes its erratic fluctuations can actually look predictable, at least in the short term. Dopamine neurons are determined to solve the flux, but most of the time there is nothing to solve. And so brain cells flail against the stochasticity, searching for lucrative patterns. Instead of seeing the randomness, we come up with imagined systems and see meaningful trends where there are only meaningless streaks. "People enjoy investing in the market and gambling in a casino for the same reason that they see Snoopy in the clouds," says the neuroscientist Read Montague. "When the brain is exposed to anything random, like a slot machine or the shape of a cloud, it automatically imposes a pattern onto the noise. But that isn't Snoopy, and you haven't found the secret pattern in the stock market."

One of Montague's recent experiments demonstrated how an unrestrained dopamine system can, over time, lead to dangerous stock-market bubbles. The brain is so eager to maximize rewards that it ends up pushing its owner off a cliff. The experiment went like this: Subjects were each given a hundred dollars and some basic information about the "current" state of the stock market. Then the players chose how much of their money to invest and nervously watched as their stock investments either rose or fell in value. The game continued for twenty rounds, and the subjects got to keep their earnings. One interesting twist was that instead of using random simulations of the stock market, Montague relied on distillations of data from history's famous markets. Montague had people "play" the Dow of 1929, the Nasdaq of 1998, the Nikkei of 1986, and the S&P 500 of 1987. This let the scientists monitor the neural responses of investors during what had once been real-life bubbles and crashes.

How did the brain deal with the fluctuations of Wall Street? The scientists immediately discovered a strong neural signal that seemed to be driving many of the investment decisions. This signal emanated from dopamine-rich areas of the brain, such as the ventral caudate, and it was encoding fictive-error learning, or the ability to learn from what-if scenarios. Take, for example, this situation: A player has decided to wager 10 percent of his total portfolio in the market, which is a rather small bet. Then he watches as the market rises dramatically in value. At this point, the fictive-error learning signal starts to appear. While he enjoys his profits, his ungrateful dopamine neurons are fixated on the profits he *missed,* as the cells compute the difference between the best possible return and the actual return. (This is a modified version of the prediction-error signal discussed earlier.) When there is a big difference between what actually happened and what might have happened—which is experienced as a feeling of regret—the player, Montague found, is more likely to do things differently the next time around. As a result, investors in the experiment adapted their investments to the ebb and flow of the market. When markets were booming, as they were in the Nasdaq bubble of the late 1990s, investors kept increasing their investments. Not to invest was to drown in regret, to bemoan all the money that might have been earned if they'd only made better decisions.

But fictive-error learning isn't always adaptive. Montague argues that these computational signals are also a main cause of financial bubbles. When the market keeps going up, people are led to make larger and larger investments in the boom. Their greedy brains are convinced that they've solved the stock market, and so they don't think about the possibility of losses. But just when investors are most convinced that the bubble isn't a bubble—many of Montague's subjects eventually put all of their money into the booming market—the bubble bursts. The Dow sinks, the Nasdaq implodes, the Nikkei collapses. All of a sudden, the same investors who'd regretted *not* fully investing in the market and had subsequently invested more were now despairing of their plummeting net worth. "You get the exact opposite effect when the market heads down," Montague says. "People just can't wait to get out, because the brain doesn't want to regret staying in." At this point, the brain realizes that it's made some very expensive prediction errors, and the investor races to dump any assets that are declining in value. That's when you get a financial panic.

The lesson here is that it's silly to try to beat the market with your brain. Dopamine neurons weren't designed to deal with the random oscillations of Wall Street. When you spend lots of money on investment-management fees, or sink your savings into the latest hot mutual fund, or pursue unrealistic growth goals, you are slavishly following your primitive reward circuits. Unfortunately, the same circuits that are so good at tracking juice rewards and radar blips will fail completely in these utterly unpredictable situations. That's why, over the long run, a randomly selected stock portfolio will beat the expensive experts with their fancy computer models. And why the vast majority of mutual funds in any given year will *underperform* the S&P 500. Even those funds that do manage to beat the market rarely do so for long. Their models work haphazardly; their successes are inconsistent. Since the market is a random walk with an upward slope, the best solution is to pick a low-cost index fund and wait. Patiently. Don't fixate on what might have been or obsess over someone else's profits. The investor who does nothing to his stock portfolio—who doesn't buy or sell a single stock—outperforms the average "active" investor by nearly 10 percent. Wall Street has always searched for the secret algorithm of financial success, but the secret is, there is no secret. The world is more random than we can imagine. That's what our emotions can't understand.

## 2

*Deal or No Deal* is one of the most popular television game shows of all time. The show has been broadcast in more than forty-five different countries, from Great Britain to Slovakia to America. The rules of the game couldn't be simpler: a contestant is confronted with twenty-six sealed briefcases each full of varying amounts of cash, from a penny to a million dollars. Without knowing the amount of money in any of the briefcases, the contestant chooses a single one, which is then placed in a lockbox. Its contents won't be revealed until the game is over.

The player then proceeds to open the remaining twenty-five briefcases one at a time. As the various monetary amounts are revealed, the contestant gradually gets an idea of how much money his or her own briefcase might contain, since all the remaining amounts are displayed on a large screen. It's a nerve-racking process of elimination, as each player tries to keep as many of the big monetary sums on the board for as long as possible. Every few rounds, a shadowy figure known as the Banker makes the player an offer for the sealed briefcase. The contestant can either accept the deal and cash out or continue to play, gambling that the unopened briefcase contains more money than the Banker has offered. As the rounds continue, the tension becomes excruciating. Spouses start crying, and children begin screaming. If the wrong briefcase is picked, or the best deal is rejected, a staggering amount of money can evaporate, just like that.

For the most part, *Deal or No Deal* is a game of dumb luck. Although players develop elaborate superstitions about the briefcases—odd numbers are better; even numbers are better; ones held by blond models are better—the monetary amounts in them are randomly distributed. There is no code to crack, no numerology to decipher. This is just fate unfolding in front of a national television audience.

And yet, *Deal or No Deal* is also a game of difficult decisions. After the Banker makes an offer, the contestant has a few minutes—usually the length of a commercial break—to make up his mind. He must weigh the prospect of sure money against the chances of winning one of the larger cash prizes. It's almost always a hard call, a moment full of telegenic anxiety.

There are two ways to make this decision. If the contestant had a calculator handy, he could quickly compare the average amount of money he might expect to win against the Banker's offer. For example, if there were three remaining briefcases, one containing $1, one containing $10,000, and one containing $500,000, then the player should, at least in theory, accept any offer over $170,000, since that is the average of the money in all three briefcases. Although offers in the early rounds are generally unfairly low—the producers don't want people to quit before it gets dramatic—as the game goes on, the offers made by the Banker become more and more reasonable, until they are essentially asymptotic with the mathematical average of the money still available. In this sense, it is extremely easy for a contestant on *Deal or No Deal* to determine whether or not to accept an offer. He just needs to add up all the remaining monetary amounts, divide that number by the number of briefcases left, and see if that figure exceeds the offer on the table. If *Deal or No Deal* were played like this, it would be a thoroughly rational game. It would also be extremely boring. It's not fun to watch people do arithmetic.

The game show is entertaining only because the vast majority of contestants don't make decisions based on the math. Take Nondumiso Sainsbury, a typical *Deal or No Deal* contestant. She is a pretty young woman from South Africa who met her husband while she was studying in America. She plans on sending her winnings back home to her poor family in Johannesburg, where her three younger brothers live in a shantytown with her mother. It's hard not to root for her to make the right decision.

Nondumiso starts off rather well. After a few rounds, she still has two big amounts—$500,000 and $400,000—left in play. As is usual for this stage of the game, the Banker makes her a blatantly unfair offer. Although the average amount of money left is $185,000, Nondumiso is offered less than half that. The producers clearly want her to keep playing.

After quickly consulting with her husband—"We still might win half a million dollars!" she shouts—Nondumiso wisely rejects the offer. The suspense builds as she prepares to pick her next briefcase. She randomly chooses a number and winces as the briefcase is slowly opened. Every second of tension is artfully mined. Nondumiso's luck has held: the briefcase contains only $300. The Banker now increases his offer to $143,000, or 75 percent of a perfectly fair offer.

After just a few seconds of deliberation, Nondumiso decides to reject the deal. Once again, the pressure builds as a briefcase is opened. The audience collectively gasps. Once again, Nondumiso has gotten lucky: she has managed to avoid eliminating either of the two big remaining sums of money. She now has a 67 percent chance of winning more than $400,000. Of course, she also has a 33 percent chance of winning $100.

For the first time, the Banker's offer is essentially fair: he is willing to "buy" Nondumiso's sealed briefcase for $286,000. As soon as she hears the number, she breaks into a huge smile and starts to cry. Without even pausing to contemplate the math, Nondumiso begins chanting, "Deal! Deal! I want a deal!" Her loved ones swarm the stage. The host tries to ask Nondumiso a few questions, and she struggles to speak through the tears.

In many respects, Nondumiso made an excellent set of decisions. A computer that meticulously analyzed the data couldn't have done much better. But it's important to note *how* Nondumiso arrived at these decisions. She never took out a calculator or estimated the average amount of money remaining in the briefcases. She never scrutinized her options or contemplated what would happen if she eliminated one of the larger amounts of money. (In that case, the offer probably would have been cut by at least 50 percent.) Instead, her risky choices were entirely impulsive; she trusted her feelings to not lead her astray.

While this instinctive decision-making strategy normally works out just fine—Nondumiso's feelings made her rich—there are certain situations on the game show that reliably fool the emotional brain. In these cases, contestants end up making terrible choices, rejecting deals that they should accept. They lose fortunes because they trust their emotions at the wrong moment.

Look at poor Frank, a contestant on the Dutch version of *Deal or No Deal*. He gets off to an unlucky start by immediately eliminating some of the most lucrative briefcases. After six rounds, Frank has only one valuable briefcase left, worth five hundred thousand euros. The Banker offers him €102,006, about 75 percent of a perfectly fair offer. Frank decides to reject the deal. He's gambling that the next briefcase he picks won't contain the last big monetary amount, thus driving up the offer from the Banker. So far, his emotions are acting in accordance with the arithmetic. They are holding out for a better deal.

But Frank makes a bad choice, eliminating the one briefcase he wanted to keep in play. He braces himself for the bad news from the Banker, who now offers Frank a deal for €2,508, or about €100,000 less than he was offered thirty seconds before. The irony is that this offer is utterly fair; Frank would be wise to cut his losses and accept the Banker's proposal. But Frank immediately rejects the deal; he doesn't even pause to consider it. After another unlucky round, the Banker takes pity on Frank and makes him an offer that's about 110 percent of the average of the possible prizes. (Tragedy doesn't make good game-show TV, and the producers are often quite generous in such situations.) But Frank doesn't want pity, and he rejects the offer. After eliminating a briefcase containing €1 — Frank's luck is finally starting to turn — he is now faced with a final decision. Only two briefcases remain: €10 and €10,000. The Banker offers him €6,500, which is a 30 percent premium over the average of the money remaining. But Frank spurns this final proposal. He decides to open his own briefcase, in the desperate hope that it contains the bigger amount. Frank has bet wrong: it contains only €10. In fewer than three minutes, Frank has lost more than €100,000.

Frank isn't the only contestant to make this type of mistake. An exhaustive analysis by a team of behavioral economists led by Thierry Post concluded that most contestants in Frank's situation act the exact same way. (As the researchers note, *Deal or No Deal* has "such desirable features that it almost appears to be designed to be an economics experiment rather than a TV show.") After the Banker's offer decreases by a large amount — this is what happened after Frank opened the €500,000 briefcase — a player typically becomes excessively risk-seeking, which means he is much more likely to reject perfectly fair offers. The contestant is so upset by the recent monetary loss that he can't think straight. And so he keeps on opening briefcases, digging himself deeper and deeper into a hole.

These contestants are victims of a very simple flaw rooted in the emotional brain. Alas, this defect isn't limited to greedy game-show contestants, and the same feelings that caused Frank to reject the fair offers can lead even the most rational people to make utterly foolish choices. Consider this scenario:

> The United States is preparing for the outbreak of an unusual Asian disease, which is expected to kill six hundred people. Two different programs to combat the disease have been proposed. Assume that the exact scientific estimates of the consequences of the programs are as follows: If program A is adopted, two hundred people will be saved. If program B is adopted, there is a one-third probability that six hundred people will be saved and a two-thirds probability that no people will be saved. Which of the two programs would you favor?

When this question was put to a large sample of physicians, 72 percent chose option A, the safe-and-sure strategy, and only 28 percent chose program B, the risky strategy. In other words, physicians would rather save a certain number of people for sure than risk the possibility that everyone might die. But consider this scenario:

> The United States is preparing for the outbreak of an unusual Asian disease, which is expected to kill six hundred people. Two different programs to combat the disease have been proposed. Assume that the exact scientific estimates of the consequences of the programs are as follows: If program C is adopted, four hundred people will die. If program D is adopted, there is a one-third probability that nobody will die and a two-thirds probability that six hundred people will die. Which of the two programs would you favor?

When the scenario was described in terms of deaths instead of survivors, physicians reversed their previous decisions. Only 22 percent voted for option C, while 78 percent chose option D, the risky strategy. Most doctors were now acting just like Frank:

they were rejecting a guaranteed gain in order to participate in a questionable gamble.

Of course, this is a ridiculous shift in preference. The two different questions examine identical dilemmas; saving one-third of the population is the same as losing two-thirds. And yet doctors reacted very differently depending on how the question was framed. When the possible outcomes were stated in terms of deaths—this is called the *loss frame*—physicians were suddenly eager to take chances. They were so determined to avoid any option associated with loss that they were willing to risk losing everything.

This mental defect—it's technical name is *loss aversion*—was first demonstrated in the late 1970s by Daniel Kahneman and Amos Tversky. At the time, they were both psychologists at Hebrew University, best known on campus for talking to each other too loudly in their shared office. But these conversations weren't idle chatter; Kahneman and Tversky (or "kahnemanandtversky," as they were later known) did their best science while talking. Their disarmingly simple experiments—all they did was ask each other hypothetical questions—helped to illuminate many of the brain's hard-wired defects. According to Kahneman and Tversky, when a person is confronted with an uncertain situation—like having to decide whether to accept an offer from the Banker—the individual doesn't carefully evaluate the information, or compute the Bayesian probabilities, or do much thinking at all. Instead, the decision depends on a brief list of emotions, instincts, and mental shortcuts. These shortcuts aren't a faster way of doing the math; they're a way of skipping the math altogether.

Kahneman and Tversky stumbled upon the concept of loss aversion after giving their students a simple survey that asked if they would accept various bets. The psychologists noticed that when a person was offered a gamble on the toss of a coin and was told that losing would cost him twenty dollars, the player demanded, on average, around forty dollars for winning. The pain of a loss was approximately twice as potent as the pleasure generated by a gain. Furthermore, decisions seemed to be determined by these feelings. As Kahneman and Tversky put it, "In human decision making, *losses loom larger than gains.*"

Loss aversion is now recognized as a powerful mental habit with widespread implications. The desire to avoid anything that smacks of loss often shapes our behavior, leading us to do foolish things. Look, for example, at the stock market. Economists have long been perplexed by a phenomenon known as the premium equity puzzle. The puzzle itself is easy to explain: over the last century, stocks have outperformed bonds by a surprisingly large margin. Since 1926, the annual return on stocks after inflation has been 6.4 percent, while the return on Treasury bills has been less than 0.5 percent. When the Stanford economists John Shoven and Thomas MaCurdy compared randomly generated financial portfolios composed of either stocks or bonds, they discovered that, over the long term, stock portfolios *always* generated higher returns than bond portfolios. In fact, stocks typically earned more than seven times as much as bonds. MaCurdy and Shoven concluded that people who invest in bonds must be "confused about the relative safety of different investments over long horizons." In other words, investors are just as irrational as game-show contestants. They, too, have a distorted sense of risk.

Classical economic theory can't explain the premium equity puzzle. After all, if investors are such rational agents, why don't all of them invest in stocks? Why are low-yield bonds so popular? In 1995, the behavioral economists Richard Thaler and Shlomo Benartzi realized that the key to solving the premium equity puzzle was loss aversion. Investors buy bonds because they hate losing money, and bonds are a safe bet. Instead of making financial decisions that reflect all the relevant statistical information, they depend on their emotional instincts and seek the

certain safety of bonds. These are well-intentioned instincts—
they prevent people from gambling away their retirement savings—but they are also misguided. The fear of losses makes investors more willing to accept a measly rate of return.

Even experts are vulnerable to these irrational feelings. Take
Harry Markowitz, a Nobel Prize–winning economist who practically invented the field of investment-portfolio theory. In the
early 1950s, while working at the RAND Corporation, Markowitz became intrigued by a practical financial question: how
much of his savings should he invest in the stock market? Markowitz derived a complicated mathematical equation that could
be used to calculate the optimal mix of assets. He had come
up with a rational solution to the old problem of risk versus
reward.

But Markowitz couldn't bring himself to use his own equation. When he divided up his investment portfolio, he ignored
the investment advice that had won him the Nobel Prize; instead
of relying on the math, he fell into the familiar trap of loss aversion and split his portfolio equally between stocks and bonds.
Markowitz was so worried about the possibility of losing his
savings that he failed to optimize his own retirement account.

Loss aversion also explains one of the most common investing mistakes: investors evaluating their stock portfolios are most
likely to sell stocks that have *increased* in value. Unfortunately,
this means that they end up holding on to their depreciating
stocks. Over the long term, this strategy is exceedingly foolish,
since ultimately it leads to a portfolio composed entirely of shares
that are losing money. (A study by Terrance Odean, an economist at UC Berkeley, found that the stocks investors sold outperformed the stocks they didn't sell by 3.4 percent.) Even professional money managers are vulnerable to this bias and tend to
hold losing stocks twice as long as winning stocks. Why does
an investor do this? Because he is afraid to take a loss—it feels

bad—and selling shares that have decreased in value makes the loss tangible. We try to postpone the pain for as long as possible; the result is more losses.

The only people who are immune to this mistake are neurologically impaired patients who can't feel any emotion at all. In most situations, these people have very damaged decision-making abilities. And yet, because they don't feel the extra sting of loss, they are able to avoid the costly emotional errors brought on by loss aversion.

Consider this experiment, led by Antonio Damasio and George Loewenstein. The scientists invented a simple investing game. In each round, the experimental subject had to decide between two options: invest $1 or invest nothing. If the participant decided not to invest, he kept the dollar, and the game advanced to the next round. If the participant decided to invest, he would hand a dollar bill to the experimenter, who would then toss a coin. Heads meant that the participant lost the $1 that was invested; tails meant that $2.50 was added to the participant's account. The game stopped after twenty rounds.

If people were perfectly rational—if they made decisions solely by crunching the numbers—then subjects would always choose to invest, since the expected overall value on each round is higher if one invests ($1.25, or $2.50 multiplied by the 50 percent chance of getting tails on the coin toss) than if one does not ($1). In fact, if a person invests on each and every round, there is a mere 13 percent chance that he'll wind up with less than twenty dollars, which is the amount a player would have if he didn't invest in any of the rounds.

So what did the subjects in Damasio's study do? Those with intact emotional brains invested only about 60 percent of the time. Because human beings are wired to dislike potential losses, most people were perfectly content to sacrifice profit for security, just like investors choosing low-yield bonds. Furthermore, the willingness of a person to invest plummeted immediately after he or she had lost a gamble — the pain of losing was too fresh.

These results are entirely predictable; loss aversion makes us irrational when it comes to evaluating risky gambles. But Damasio and Loewenstein didn't stop there. They also played the investing game with neurologically impaired patients who could no longer experience emotion. If it was the feeling of loss aversion that caused these bad investing decisions, then these patients should perform *better* than their healthy peers.

That's exactly what happened. The emotionless patients chose to invest 83.7 percent of the time and gained significantly more money than normal subjects. They also proved much more resistant to the misleading effects of loss aversion, and they gambled 85.2 percent of the time after a lost coin toss. In other words, losing money made them *more* likely to invest as they realized that investing was the best way to recoup their losses. In this investing situation, having no emotions was a crucial advantage.

And then there is *Deal or No Deal,* which turns out to be a case study in loss aversion. Imagine you are Frank. Less than a minute ago, you turned down the Banker's offer of €102,006. But now you've picked the worst possible briefcase, and the offer has declined to €2,508. In other words, you've lost a cool hundred grand. Should you accept the current deal? The first thing your mind does is make a list of the options under consideration. However, instead of evaluating those options in terms of arithmetic — which would be the rational thing to do — you use your emotions as a shortcut to judgment. You simulate the various scenarios and see how each makes you feel. When you imagine accepting the offer of €2,508, you experience a sharply negative emotion, even though it's a perfectly fair offer. The problem is that your emotional brain interprets the offer as a dramatic loss, since it's automatically compared to the much larger amount of money that had been on offer just a few moments earlier. This resulting feeling serves as a signal that accepting the deal is a bad idea; you should reject the offer and open another briefcase. In this situation, loss aversion makes you risk seeking.

But now that you've imagined rejecting the offer, you fixate on the highest monetary amount now possible. This is the potential gain you measure everything against, what economists call the reference point. (For Frank, the potential gain during the final rounds was €10,000. For the physicians being quizzed about that unusual Asian disease, the potential gain was saving all six hundred people.) When you think about this optimistic possibility, you experience, however briefly, a pleasurable feeling. You contemplate the upside of risk and envision a check with lots of zeros. You might not be able to get back the €100,000 offer, but at least you won't leave empty-handed.

The upshot of all this is that you badly miscalculate the risk. You keep on chasing after the possibility of a big gain because you can't accept the prospect of a loss. Your emotions have sabotaged common sense.

Loss aversion is an innate flaw. Everyone who experiences emotion is vulnerable to its effects. It's part of a larger psychological phenomenon known as negativity bias, which means that, for the human mind, *bad is stronger than good*. This is why in marital interactions, it generally takes at least five kind comments to compensate for one critical comment. As Jonathan Haidt points out in his book *The Happiness Hypothesis*, people believe that a person who's been convicted of murder must perform at least twenty-five acts of "life-saving heroism" in order to make up for his or her one crime. There's no rational reason for us to treat gains and losses or compliments and criticisms so differently. But we do. The only way to avoid loss aversion is to know about the concept.

# 3

"The credit card is my enemy," Herman Palmer says. Herman is a very friendly guy, with sympathetic eyes and a wide smile that fills his face, but when he starts to talk about credit cards, his demeanor abruptly darkens. He furrows his brow, lowers his voice, and leans forward in his chair. "Every day, I see lots of smart people who have the same problem: Visa and MasterCard. Their problem is all those plastic cards they've got in their wallet." Then he shakes his head in dismay and lets out a resigned sigh.

Herman is a financial counselor in the Bronx. He has spent the last nine years working for GreenPath, a nonprofit organization that helps people deal with their debt problems. His small office is a minimalist affair, with a desk so clean that it looks as if no one has ever used it. The only thing on the desk is a large glass candy jar, but this jar isn't stuffed with M&M's or jelly beans or miniature candy bars. It's filled with the cut-up shards of hundreds of credit cards. The plastic pieces make for a pretty collage—the iridescent security stickers glitter in the light—but Herman doesn't keep the jar around for aesthetic reasons. "I use it as a kind of shock treatment," he says. "I'll ask a client for their cards and just cut them up right in front of them. And then I just add the cards to the jar. I want people to see that they are not alone, that so many people have the exact same problem." Once the jar in his office is completely filled—and that only takes a few months—Herman empties it into the big glass vase in the waiting room. "That's our flower display," he jokes.

According to Herman, the jar of credit cards captures the essence of his job. "I teach people how *not* to spend money," he says. "And it's damn near impossible to not spend money if you've still got all these cards, which is why I always cut them up." The first time I visited the GreenPath office was a few weeks after Christmas, and the waiting room was full of anxious-looking people trying to pass the time with old issues of celebrity magazines. Every chair was taken. "January is our busiest time of year," Herman says. "People always overspend during the holidays, but they don't realize how much they've overspent until the credit card bills arrive in the mail. That's when they come see us."

For the most part, Herman's clients are from the neighborhood, a working-class area of row houses that were once single-family dwellings but are now apartment buildings, with numerous buzzers and mailboxes grafted onto the front doors. Many of the homes have fallen into disrepair, with peeling siding and graffiti. There aren't any supermarkets nearby, but there are plenty of bodegas and liquor stores. A little farther down the block, there are two pawnshops and three check-cashing operations. Every few minutes, another number 6 subway train rumbles directly overhead, shrieking to a stop near the GreenPath office. It's the last stop on the line.

Nearly half of Herman's clients are single mothers. Many of these women work full-time but still struggle to pay their bills. Herman estimates that his clients spend, on average, around 40 percent of their income on housing, even though the neighborhood has some of the cheapest real estate in New York City. "It's easy to judge people," Herman says. "It's easy to think, 'I would never have gotten into so much debt,' or to think that just because someone needs financial help, then they must be irresponsible. But a lot of the people I see are just trying to make ends meet. The other day I had a mother come in who just broke my heart. She was working two jobs. Her credit card bill was all daycare charges for her kid. What am I supposed to tell her? That her kid can't go to daycare?"

This ability to help his clients without judging them, to understand what they're going through, is what makes Herman such a good financial counselor. (He has an unusually high success rate, with more than 65 percent of his clients completing their debt-elimination plans.) It would be easy for Herman to play the scold, to chastise his clients for letting their spending get out of control. But he does just the opposite. Instead of lecturing his clients, he listens to them. After Herman destroys their credit cards at the initial meeting—he almost always gets out his scissors within the first five minutes—he will spend the next several hours poring over their bills and bank statements, trying to understand what's gone wrong with their finances. Is their rent too expensive? Are they spending too much money on clothes or cell phones or cable television? "I always tell my clients that they are going to leave my office with a practical plan," Herman says. "And charging it to Mr. MasterCard is not a plan."

When Herman talks about the people who have been helped by his financial advice, his face takes on the glow of a proud parent. There's the plumber from Co-op City who lost his job and started paying rent with his credit card. After a few months, his interest rate was above 30 percent. Herman helped him consolidate his debt and get his expenses under control. There's that single mother who couldn't afford daycare. "We helped her find other ways to save money," he says. "We cut her expenses by enough so that she didn't have to charge everything. The trick is to notice whenever you're spending money. All that little stuff? Guess what: it adds up." There's the schoolteacher who racked up debt on ten different credit cards and paid hundreds of dollars every month in late fees alone. It took five years of careful discipline, but now the teacher is debt free. "I know the client is going to be okay when they start telling me about the sweater or CD they really wanted but they didn't buy," Herman says. "That's when I know they are starting to make better decisions."

Most of the people who come to see Herman tell the same basic story. One day, a person gets a credit card offer in the mail. (Credit card companies sent out 5.3 billion solicitations in 2007, which means the average American adult got fifteen offers.) The card seems like such a good deal. In big bold print it advertises a low introductory rate along with something about getting cash back or frequent-flier miles or free movie tickets. And so the person signs up. He fills out the one-page form and then, a few weeks later, gets a new credit card in the mail. At first, he doesn't use it much. Then one day he forgets to get cash, and so he uses the new credit card to pay for food at the supermarket. Or maybe the refrigerator breaks, and he needs a little help buying a new one. For the first few months, he always manages to pay off the full bill. "Almost nobody gets a credit card and says, 'I'm going to use this to buy things I can't afford,'" Herman says. "But it rarely stays like that for long."

According to Herman, the big problem with credit cards — the reason he enjoys cutting them up so much — is that they cause people to make stupid financial choices. They make it harder to resist temptation, so people spend money they don't have. "I've seen it happen to the most intelligent people," Herman says. "I'll look at their credit card bill and I'll see a charge for fifty dollars at a department store. I'll ask them what they bought. They'll say, 'It was a pair of shoes, Herman, but it was on sale.' Or they'll tell me that they bought another pair of jeans but the jeans were fifty percent off. It was such a good deal that it would have been dumb *not* to buy it. I always laugh when I hear that one. I then have them add up all the interest they are going to pay on those jeans or that pair of shoes. For a lot of these people, it will be around twenty-five percent a month. And you know what? Then it's not such a good deal anymore."

These people aren't in denial. They know they have serious debt problems and that they're paying a lot of interest on their debts. That's why they're visiting a financial adviser. And yet, they *still* bought the jeans and the pair of shoes on sale. Herman is all too familiar with the problem: "I always ask people, 'Would you have bought the item if you had to pay cash? If you had to go to an ATM and feel the money in your hands and then hand it over?' Most of the time, they think about it for a minute and then they say no."

Herman's observations capture an important reality about credit cards. Paying with plastic fundamentally changes the way we spend money, altering the calculus of our financial decisions. When you buy something with cash, the purchase involves an actual loss—your wallet is literally lighter. Credit cards, however, make the transaction abstract, so that you don't really feel the downside of spending money. Brain-imaging experiments suggest that paying with credit cards actually reduces activity in the insula, a brain region associated with negative feelings. As George Loewenstein, a neuroeconomist at Carnegie Mellon, says, "The nature of credit cards ensures that your brain is anesthetized against the pain of payment." Spending money doesn't feel bad, so you spend more money.

Consider this experiment: Drazen Prelec and Duncan Simester, two business professors at MIT, organized a real-life, sealed-bid auction for tickets to a Boston Celtics game. Half the participants in the auction were informed that they had to pay with cash; the other half were told they had to pay with credit cards. Prelec and Simester then averaged the bids for the two different groups. Lo and behold, the average credit card bid was *twice* as high as the average cash bid. When people used their Visas and MasterCards, their bids were much more reckless. They no longer felt the need to contain their expenses, and so they spent way beyond their means.

This is what's happened to the American consumer over the past few decades. The statistics are bleak: the average household currently owes more than nine thousand dollars in credit card debt, and the average number of credit cards per person is 8.5. (More than 115 million Americans carry month-to-month balances on their credit cards.) In 2006, consumers spent more than seventeen billion dollars in penalty fees alone on their credit cards. Since 2002, Americans have had a negative savings rate, which means that we've spent more than we've earned. The Federal Reserve recently concluded that this negative savings rate was largely a consequence of credit card debt. We spend so much money on interest payments that we can't save for retirement.

At first glance, this behavior makes no sense. Given the exorbitant interest rates charged by most credit card companies —rates of 25 percent or more are common—a rational consumer would accumulate credit card debt only as a last resort. Paying interest is expensive. And yet, credit card debt is as American as apple pie. "The people who have credit card debt are the same people who drive an extra mile to save two cents on a gallon of gas," Herman says. "They are the same people who clip coupons and comparison shop. Many of these people are normally very good with their money. But then they bring me their credit card bill and they say, 'I don't know what happened. I don't know how I spent all this money.'"

The problem with credit cards is that they take advantage of a dangerous flaw built into the brain. This failing is rooted in our emotions, which tend to overvalue immediate gains (like a new pair of shoes) at the cost of future expenses (high interest rates). Our feelings are thrilled by the prospect of an immediate reward, but they can't really grapple with the long-term fiscal consequences of that decision. The emotional brain just doesn't understand things like interest rates or debt payments or finance charges. As a result, areas like the insula don't react to transactions involving a Visa or MasterCard. Because our impulsivity encounters little resistance, we swipe our cards and buy whatever we want. We'll figure out how to pay for it later.

This sort of shortsighted decision-making isn't dangerous only for people with too many credit cards in their wallets. In recent years, Herman has seen a new financial scourge in the neighborhood: subprime mortgages. "I still remember the first subprime mortgage I dealt with," Herman says. "I remember thinking, 'This is a really bad deal. These people just bought a house that's way too expensive for them, and they don't even know it yet.' And that's when I knew that I'd be seeing a lot of these loans in the future."

The most common type of subprime mortgage that Herman deals with is the 2/28 loan, which comes with a low, fixed-interest rate for the first two years and a much higher, adjustable rate for the next twenty-eight. In other words, the loan works a lot like a credit card: it lets people get homes for virtually nothing up front, then hits the borrowers with high-interest payments at some point in the distant future. By the time the housing market went bust in the summer of 2007, subprime loans like the 2/28 accounted for almost 20 percent of all mortgages. (The percentage in poorer neighborhoods, such as the Bronx, was much higher, with more than 60 percent of all mortgages falling into the subprime category.) Unfortunately, the loan comes with a steep cost. The structure of the loan ensures that subprime borrowers are five times more likely to default than other borrowers. Once the rates start to rise—and they always do—many people can no longer afford the monthly mortgage payments. By the end of 2007, a whopping 93 percent of completed foreclosures involved adjustable-rate loans that had recently been adjusted. "When I help people with a mortgage," Herman says, "I never ask them about the home. Because then they just start talking about how pretty it is and how the extra room will be so great for their kids. That's just temptation talking. I make sure we stick to the numbers and that we especially focus on their interest payments in the future, after the rates are adjusted." While 2/28 loans tempt consumers with low initial payments, that temptation turns out to be extremely expensive. In fact, subprime loans even proved tempting for people with credit scores that qualified them for conventional loans that had far better financial terms. During the peak of the housing boom, 55 percent of all 2/28 mortgages were sold to homeowners who could have gotten prime mortgages. Although prime mortgages would have saved them lots of money over the long term, these people just couldn't resist the allure of those low initial payments. Their feelings tricked them into making foolish financial decisions.

The pervasive reach of credit cards and subprime loans reveals our species' irrationality. Even when people are committed to long-term goals, such as saving for retirement, they are led astray by momentary temptations. Our impulsive emotions make us buy what we can't afford. As Plato might have put it, the horses are pulling the charioteer against his will.

Understanding the circuitry of temptation is one of the practical ambitions of scientists studying decision-making. Jonathan Cohen, a neuroscientist at Princeton University, has made some important progress. He's begun to diagnose the specific brain regions responsible for the attraction to credit cards and subprime loans. One of his recent experiments involved putting a subject in an fMRI machine and making him decide between a small Amazon gift certificate that he could have right away and a slightly larger gift certificate that he'd receive in two to four weeks. Cohen discovered that these two options activated very different neural systems. When a subject contemplated a gift certificate in the future, brain areas associated with rational planning, such as the prefrontal cortex, were more active. These cortical regions urge a person to be patient, to wait a few extra weeks for the bigger gain.

However, when a subject started thinking about getting the gift certificate right away, the brain areas associated with emotion—such as the midbrain dopamine system and nucleus accumbens—were turned on. These are the cells that tell a person to take out a mortgage he can't afford, or run up credit card debt when he should be saving for retirement. All these cells want is a reward, and they want it now.

By manipulating the amount of money on offer in each situation, Cohen and his collaborators could watch this neural tug of war unfold. They saw the fierce argument between reason and feeling as the mind was pulled in contradictory directions. The ultimate decision—whether to save for the future or to indulge in the present—was determined by whichever region showed greater activation. The people who couldn't wait for the bigger gift certificates—and most people couldn't wait—were led astray by their feelings. More emotions meant more impulsivity. (This also helps explain why men who are shown revealing pictures of attractive women, what scientists refer to as "reproductively salient stimuli," become even more impulsive: the photos activate their emotional circuits.) However, subjects who chose to wait and receive the larger Amazon gift certificates later showed increased activity in their prefrontal cortices; they did the math and selected the "rational" option.

This discovery has important implications. For starters, it locates the neural source for many financial errors. When self-control breaks down, and we opt for the rewards we can't afford, it's because the rational brain has lost the neural tug of war. David Laibson, an economist at Harvard and coauthor of the paper on the monetary-reward experiment, notes: "Our emotional brain wants to max out the credit card, order dessert, and smoke a cigarette. When it sees something it wants, it has difficulty waiting to get it." Corporations have learned to take advantage of this limbic impatience. Consider the teaser rates offered in credit card solicitations. In order to entice new customers, lenders typically advertise their low introductory charges. These alluring offers expire within a few short months, leaving customers stuck with lots of debt on credit cards with high interest rates. The bad news is that the emotional brain is routinely duped by these tempting (but financially foolish) advertisements. "I always tell people to read *only* the fine print," Herman says. "The bigger the print, the less it matters."

Unfortunately, most people don't follow Herman's advice. Lawrence Ausubel, an economist at the University of Maryland, analyzed the responses of consumers to two different credit card promotions used by actual credit card companies. The first card offered a six-month teaser rate of 4.9 percent that was followed by a lifetime at 16 percent. The second card had a slightly higher teaser rate — 6.9 percent — but a significantly lower lifetime rate (14 percent). If consumers were rational, they would always choose the card with the lower lifetime rate, since that's the rate that would apply to most of their debts. Of course, this isn't what happens. Ausubel found that the credit offer with the 4.9 percent teaser rate was chosen by consumers almost three times more often than the other. Over the long term, this impatience leads to significantly higher interest payments.

When people opt for bad credit cards, or choose 2/28 mortgages, or fail to put money in their 401(k)s, they are acting just like the experimental subjects who chose the wrong Amazon gift certificate. Because the emotional parts of the brain reliably undervalue the future — life is short and we want pleasure *now* — we all end up spending too much money today and delaying saving until tomorrow (and tomorrow and tomorrow). George Loewenstein, the neuroeconomist, thinks that understanding the errors of the emotional brain will help policymakers develop plans that encourage people to make better decisions: "Our emotions are like software programs that evolved to solve important and recurring problems in our distant past," he says. "They are not always well suited to the decisions we make in modern life. It's important to know how our emotions lead us astray so that we can find ways to compensate for these flaws."

Some economists are already working on that. They are using this brain-imaging data to support a new political philosophy known as asymmetric paternalism. That's a fancy name for a simple idea: creating policies and incentives that help people triumph over their irrational impulses and make better, more prudent decisions. Shlomo Benartzi and Richard Thaler, for example, designed a 401(k) that takes our irrationality into account.

Their plan, called Save More Tomorrow, neatly sidesteps the limbic system. Instead of asking people if they want to start saving right away — which is the standard pitch for a 401(k) — companies in the Save More Tomorrow program ask their employees if they want to opt into savings plans that begin in a few months. Since this proposal allows people to make decisions about the future without contemplating possible losses in the present, it bypasses their impulsive emotional brains. (This is roughly equivalent to asking a person if he wants a ten-dollar Amazon gift certificate in one year or an eleven-dollar gift certificate in one year and one week. In this case, virtually everyone chooses the rational option, which is the larger amount.) Trial studies of this program show it's a resounding success: after three years, the average savings rate has gone from 3.5 percent to 13.6 percent.

Herman is content with an even simpler solution. "My first piece of advice is always the same," he says. "Cut up the damn cards. Or put them in a block of ice in the freezer. Learn to pay with cash." Herman knows from experience that unless people get rid of their credit cards, they won't be able to stay on fiscally sound spending plans: "I've seen people who have more debt than you can believe, and they'll still make irresponsible shopping decisions if they can charge it." It's not easy for the brain to choose a long-term gain over an immediate reward — such a decision takes cognitive effort — which is why getting rid of anything that makes the choice harder (such as credit cards) is so important. "Everybody knows about temptation," Herman says. "Everybody wants that new pair of shoes and the big house. But sometimes you have to say no to yourself." He tries to quote a famous song by the Rolling Stones but he can't quite remember the lyrics. The message of the chorus is simple: you can't always get what you want, but sometimes *not* getting what you want is just what you need.

# 4

# The Uses of Reason

The summer of 1949 had been long and dry in Montana; the grassy highlands were like tinder. On the afternoon of August 5 — the hottest day ever recorded in the area — a stray bolt of lightning set the ground on fire. A parachute brigade of firefighters, known as smokejumpers, was dispatched to put out the blaze. Wag Dodge, a veteran with nine years of smokejumping experience, was in charge. When the jumpers took off from Missoula in a C-47, a military transport plane left over from World War II, they were told that the fire was small, just a few burning acres in the Mann Gulch river valley. As the plane approached the fire, the jumpers could see the smoke in the distance. The hot wind blew it straight across the sky.

Mann Gulch is a place of geological contradiction. It is where the Rocky Mountains meet the Great Plains, pine trees give way to prairie grass, and the steep cliffs drop onto the steppes of the Midwest. The gulch is just over three miles long, but it marks the border between these two different terrains.

The fire began on the Rockies' side, on the western edge of the gulch. By the time the firefighters arrived at the gulch, the blaze had grown out of control. The surrounding hills had all been burned; the landscape was littered with the skeletons of pine trees. Dodge moved his men over to the grassy side of the gulch and told them to head downhill, toward the placid Missouri River. Dodge didn't trust this blaze. He wanted to be near water; he knew this fire could crown.

Crowns occur when flames get so high they reach into the top branches of trees. Once that happens, the fire has too much fuel. Hot embers begin to swirl in the air, spreading the fire across the prairie. The smokejumpers used to joke that the only way to control a crown fire was to pray like hell for rain. Norman Maclean, in his seminal history *Young Men and Fire,* described what it was like to be close to such a fire:

> It sounds like a train coming too fast around a curve and may get so high-keyed that the crew cannot understand what their foreman is trying to do to save them. Sometimes, when the timber thins out, it sounds as if the train were clicking across a bridge, sometimes it hits an open clearing and becomes hushed as if going through a tunnel, but when the burning cones swirl through the air and fall on the other side of the clearing, the new fire sounds as if it were the train coming out of the tunnel, belching black unburned smoke. The unburned smoke boils up until it reaches oxygen, then bursts into gigantic flames on top of its cloud of smoke in the sky. The new [novice] firefighter, seeing black smoke rise from the ground and then at the top of the sky turn into flames, thinks that natural law has been reversed.

Dodge looked at the dry grass and the dry pine needles. He felt the hot wind and the hot sun. The conditions were making him nervous. To make matters worse, the men had no map of the terrain. They were also without a radio, since the parachute on the radio pack had failed to open and the transmitter had been smashed on the rocks. The small crew of smokejumpers was all alone with this fire; there was nothing between them and it but a river and a thick tangle of ponderosa pine and Douglas fir trees. And so the jumpers set down their packs and watched the blaze from across the canyon. When the wind parted the smoke, as it did occasionally, they could see inside the fire as the flames leaped from tree to tree.

It was now five o'clock—a dangerous time to fight wilderness fires because the twilight wind can shift without warning. The breeze had been blowing the flames up the canyon, away from the river. But then, suddenly, the wind reversed. Dodge saw the ash swirl in the air. He saw the top of the flames flicker and wave. And then he saw the fire leap across the gulch and spark the grass on his side.

That's when the updraft began. Fierce winds began to howl through the canyon, blowing straight toward the men. Dodge could only watch as the fire became an inferno. He was suddenly staring at a wall of flame two hundred feet tall and three hundred feet deep on the edge of the prairie. In a matter of seconds, the flames began to devour the grass on the slope. The fire ran toward the smokejumpers at thirty miles per hour, incinerating everything in its path. At the fire's center, the temperature was more than two thousand degrees, hot enough to melt rock.

Dodge screamed at his men to retreat. It was already too late to run to the river, since the fire was blocking their path. Each man dropped his fifty pounds of gear and started running up the brutally steep canyon walls, trying to get to the top of the ridge and escape the blowup. Because heat rises, a fire that starts burning on flat prairie accelerates when it hits a slope. On a 50 percent grade, a fire will move nine times faster than it does on level land. The slopes at Mann Gulch are 76 percent.

When the fire first crossed the gulch, Dodge and his crew had a two-hundred-yard head start. After a few minutes of running, Dodge could feel the fierce heat on his back. He glanced over his shoulder and saw that the fire was now fewer than fifty yards away and gaining. The air began to lose its oxygen. The fire was sucking the wind dry. That's when Dodge realized the blaze couldn't be outrun. The hill was too steep, and the flames were too fast.

So Dodge stopped running. He stood perfectly still as the fire accelerated toward him. Then he started yelling at his men to do the same. He knew they were racing toward their own immolation and that in fewer than thirty seconds the fire would run them over, like a freight train without brakes. But nobody stopped. Perhaps the men couldn't hear Dodge over the deafening roar of flames. Or perhaps they couldn't bear the idea of stopping. When confronted with a menacing fire, the most basic instinct is to run away. Dodge was telling the men to stand still.

But Dodge wasn't committing suicide. In a fit of desperate creativity, he came up with an escape plan. He quickly lit a match and ignited the ground in front of him. He watched as those flames raced away from him, up the canyon walls. Then Dodge stepped into the ashes of this smaller fire, so that he was surrounded by a thin buffer of burned land. He lay down on the still smoldering embers. He wet his handkerchief with some water from his canteen and clutched the cloth to his mouth. He closed his eyes tight and tried to inhale the thin ether of oxygen remaining near the ground. Then he waited for the fire to pass around him. After several terrifying minutes, Dodge emerged from the ashes virtually unscathed.

Thirteen smokejumpers were killed by the Mann Gulch fire. Only two men in the crew besides Dodge managed to survive, and that was because they found a shallow crevice in the rocky hillside. As Dodge had predicted, the flames were almost impossible to outrun. White crosses still mark the spots where the men died; all of the crosses are below the ridge.

# 1

Dodge's escape fire is now a standard firefighting technique. It has saved the lives of countless firefighters trapped by swift blazes. At the time, however, Dodge's plan seemed like sheer madness. His men could think only about fleeing the flames, and yet their leader was starting a new fire. Robert Sallee, a first-year smokejumper who survived the blaze, later said he'd thought that "Dodge had gone nuts, just plain old nuts."

But Dodge was perfectly sane. In the heat of the moment he managed to make a very smart decision. The question, for those of us looking back on it, is how? What allowed him to resist the urge to flee? Why didn't he follow the rest of his crew up the gulch? Part of the answer is experience. Most of the smokejumpers were teenagers working summer jobs. They had fought only a few fires, and none of them had ever seen a fire like that. Dodge, on the other hand, was a grizzled veteran of the forest service; he knew what prairie flames were capable of. Once the fire crossed the gulch, Dodge realized that it was only a matter of time before the men were caught by the hungry flames. The slopes were too steep and the wind was too fierce and the grass was too dry; the blaze would beat them to the top. Besides, even if the men managed to reach the top of the mountain, they were still trapped. The ridge was covered with high, dry grass that hadn't been trimmed by cattle. It would burn in an instant.

For Dodge, it must have been a moment of unspeakable horror: to know that there was nowhere to go; to realize that his men were running to their deaths and that the wall of flame would consume them all. But Dodge's fear wasn't what saved him. In fact, the overwhelming terror of the situation was part of the problem. After the fire started burning uphill, all of the smokejumpers became fixated on getting to the ridge, even though the ridge was too far away for them to reach. Walter Rumsey, a first-year smokejumper, later recounted what was going through his mind when he saw Dodge stop running and get out his matchbook. "I remember thinking that that was a very good idea," Rumsey said, "but I don't remember what I thought it was good for . . . I kept thinking the ridge—if I can make it. On the ridge I will be safe." William Hellman, the second in command, looked at Dodge's escape fire and reportedly said, "To hell with that, I'm getting out of here." (Hellman did reach the ridge, the only smokejumper who managed to do so, but he died the next day from third-degree burns that covered his entire body.) The rest of the men acted the same way. When Dodge was asked during the investigation why none of the smokejumpers followed his orders to stop running, he just shook his head. "They didn't seem to pay any attention," he said. "That is the part I didn't understand. They seemed to have something on their minds—all headed in one direction . . . They just wanted to get to the top."

Dodge's men were in the grip of panic. The problem with panic is that it narrows one's thoughts. It reduces awareness to the most essential facts, the most basic instincts. This means that when a person is being chased by a fire, all he or she can think about is running from the fire.

This is known as perceptual narrowing. In one study, people were put one at a time in a pressure chamber and told that the pressure would slowly be increased until it simulated that of a sixty-foot dive. While inside the pressure chamber, the subject was asked to perform two simple visual tasks. One task was to respond to blinking lights in the center of the subject's visual field, and the other involved responding to blinking lights in his peripheral vision. As expected, each of the subjects inside the pressure chamber exhibited all the usual signs of panic—a racing pulse, elevated blood pressure, and a surge of adrenaline. These symptoms affected performance in a very telling way.

Although the people in the pressure chamber performed just as well as control subjects did on the central visual task, those in the pressure chamber were twice as likely to miss the stimuli in their peripheral vision. Their view of the world literally shrank.

The tragedy of Mann Gulch holds an important lesson about the mind. Dodge survived the fire because he was able to beat back his emotions. Once he realized that his fear had exhausted its usefulness — it told him to run, but there was nowhere to go — Dodge was able to resist its primal urges. Instead, he turned to his conscious mind, which is uniquely capable of deliberate and creative thought. While automatic emotions focus on the most immediate variables, the rational brain is able to expand the list of possibilities. As the neuroscientist Joseph LeDoux says, "The advantage [of the emotional brain] is that by allowing evolution to do the thinking for you at first, you basically buy the time that you need to think about the situation and do the most reasonable thing." And so Dodge stopped running. If he was going to survive the fire, he needed to think.

What Dodge did next relied entirely on the part of his brain that he could control. In the panic of the moment, he was able to come up with a new solution to his seemingly insurmountable problem. There was no pattern to guide him — no one had ever started an escape fire before — but Dodge was able to imagine his survival. In that split second of thought, he realized that it was possible to start his own fire, and that this fire might give him a thin barrier of burned earth. "It just seemed like the logical thing to do," Dodge said. He didn't know if his escape fire would work — he thought he would probably suffocate — but it still appeared to be a better idea than running. And so Dodge felt for the direction of the wind and lit the prairie weeds right in front of him. They ignited like paper. The surrounding tinder wilted to ash. He had made a firewall out of fire.

This kind of thinking takes place in the prefrontal cortex, the outermost layer of the frontal lobes.* Pressed tight against the bones of the forehead, the prefrontal cortex has undergone a dramatic expansion in the human brain. When you compare a modern human cortex to that of any other primate, or even to some of our hominid ancestors', the most obvious anatomical difference is this swelling at the fore. The Neanderthal, for example, had a slightly larger brain than *Homo sapiens*. But he still had the prefrontal cortex of a chimp. As a result, Neanderthals were missing one of the most important talents of the human brain: rational thought.

*Rationality* can be a difficult word to define—it has a long and convoluted intellectual history—but it's generally used to describe a particular style of thinking. Plato associated rationality with the use of logic, which he believed made humans think like the gods. Modern economics has refined this ancient idea into rational-choice theory, which assumes that people make decisions by multiplying the *probability* of getting what they want by the *amount of pleasure* (utility) that getting what they want will bring. This reasonable rubric allows us all to maximize our happiness, which is what rational agents are always supposed to do.

Of course, the mind isn't a purely rational machine. You don't compute utility in the supermarket or use math when throwing a football or act like the imaginary people in economics textbooks. The Platonic charioteer is often trounced by his emotional horses. Nevertheless, the brain does have a network of rational parts, centered in the prefrontal cortex. If it weren't for these peculiar lumps of gray and white matter, we couldn't even conceive of rationality, let alone act in a rational manner.

---

*Although certain sections of this brain area, such as the orbitofrontal cortex, are actually concerned with the perception of emotional states, the upper two-thirds of the prefrontal cortex—particularly the dorsolateral prefrontal cortex, or DLPFC—is generally regarded as the rational center of the brain. When you crunch numbers, deploy logic, or rely on deliberate analysis, you're using your DLPFC.

The prefrontal cortex was not always held in such high regard. When scientists first began dissecting the brain in the nineteenth century, they concluded that the frontal lobes were useless, unnecessary folds of flesh. Unlike other cortical areas, which were responsible for specific tasks such as controlling the body or generating language, the prefrontal cortex seemed to do nothing. It was the appendix of the mind. As a result doctors figured they might as well find out what happened when the area was excised. In 1935, the Portuguese neurologist Antonio Egas Moniz performed the first prefrontal leucotomy, a delicate surgery during which small holes were cut into frontal lobes. (The surgery was inspired by reports that chimpanzees became less aggressive after undergoing similar procedures.) Moniz restricted the surgery to patients with severe psychiatric problems, such as schizophrenia, who would otherwise be confined to dismal mental institutions. The leucotomy certainly wasn't a cure-all, but many of Moniz's patients did experience a reduction in symptoms. In 1949, he was awarded the Nobel Prize in Medicine for pioneering the procedure.

The success of the leucotomy led doctors to experiment with other kinds of frontal lobe operations. In the United States, Walter Freeman and James Watts developed a procedure known as the prefrontal lobotomy, which was designed to completely ablate the tracts of white matter connecting the prefrontal cortex and the thalamus. The surgery was brutally simple: a thin blade was inserted just under the eyelid, hammered through a thin layer of bone, and shimmied from side to side. The treatment quickly became exceedingly popular. Between 1939 and 1951, the "cutting cure" was performed on more than eighteen thousand patients in American asylums and prisons.

Unfortunately, the surgery came with a wide range of tragic side effects. Between 2 and 6 percent of all patients died on the operating table. Those who survived were never the same. Some patients sank into a stupor, utterly uninterested in everything around them. Others lost the ability to use language. (This is what happened to Rosemary Kennedy, the sister of President John F. Kennedy. Her lobotomy was given as a treatment for "agitated depression.") The vast majority of lobotomized patients suffered from short-term memory problems and the inability to control their impulses.

The frontal lobe lobotomy, unlike Moniz's leucotomy, was a crude procedure. Its path of destruction was haphazard and unpredictable. Although doctors tried to cut only the connections to the prefrontal cortex, they really didn't know what they were cutting. However, over the past several decades, neurologists have studied this brain area with great precision. They now know exactly what happens when the prefrontal cortex is damaged.

Consider the case of Mary Jackson, an intelligent and driven nineteen-year-old with a bright future. Although she grew up in a blighted inner-city neighborhood, Mary received a full scholarship to an Ivy League university. She was a history major with a pre-med concentration and hoped one day to become a pediatrician so she could open up a medical clinic in her old neighborhood. Her boyfriend, Tom, was an undergraduate at a nearby college, and they planned to get married after Mary finished medical school.

But then, in the summer after her sophomore year, Mary's life began falling apart. Tom noticed it first. Mary had never drunk alcohol before—her parents were strict Baptists—but she suddenly started frequenting bars and dance clubs. She began sleeping with random men and experimented with crack cocaine. She disowned her old friends, avoided church, and broke up with Tom. Nobody knew what had gotten into her.

When Mary returned to school, her grades began to slip. She stopped attending class. Her semester report card was dismal: three F's and two D's. Mary's adviser warned her that she would lose her scholarship and recommended psychiatric counseling. But Mary ignored the suggestion and continued to spend most of her nights at the local bar.

Later that spring, Mary developed a high fever and a hacking cough. At first, she assumed her sickness was just the side effect of too much partying, but the sickness wouldn't go away. She went to the student health center and was diagnosed with pneumonia. But even after she was treated with intravenous antibiotics and oxygen, the fever lingered. Mary's immune system seemed compromised. The doctors ordered more blood tests. That's when Mary learned she was HIV-positive.

Mary immediately broke down in a fit of hysterical tears. She told her doctor that she didn't understand her own behavior. Until the previous summer, she had never felt the urge to do drugs or sleep around or skip class. She had been diligently focused on her long-term goals of going to medical school and starting a family with Tom. But now she was unable to control her own impulses. She couldn't resist temptation. She made one reckless decision after another.

Mary's doctor referred her to Dr. Kenneth Heilman, a distinguished neurologist now at the University of Florida. Heilman began by giving Mary some simple psychological tests: He asked her to remember a few different objects, and then distracted her for thirty seconds by having her count backward. When Heilman asked Mary if she could still remember the objects, she looked at him with a blank stare. Her working memory had vanished. When Heilman tried to give Mary a different memory test, she flew into a rage. He asked her if she had always had such a bad temper. "Up to about a year ago, it was extremely rare that I got angry," Mary said. "Now it seems I am always flying off the handle."

All of these neurological symptoms—the diminished memory capacity, the self-destructive impulsiveness, the uncontrollable rage—suggested a problem with Mary's prefrontal cortex.

So Heilman gave Mary a second round of tests: He put a comb in front of her but told her not to touch it. She immediately started combing her hair. He put a pen and paper in front of her but told her to keep her hands still. She automatically started writing. After scribbling a few sentences, however, Mary became bored and started looking for a new distraction. "It seemed that rather than having internal goals motivate her behavior," Heilman wrote in his clinical report, "she was entirely stimulus dependent." Whatever Mary saw, she touched. Whatever she touched, she wanted. Whatever she wanted, she needed.

Heilman ordered an MRI. That's when he saw the tumor: a large mass emanating from the pituitary gland and pressing on Mary's prefrontal cortex. This was the cause of her decline. That growth had left her with executive dysfunction, an inability to maintain a coherent set of goals and contemplate the consequences of one's actions. As a result, Mary was unable to act on any ideas but the most immediate. The tumor had erased some of the necessary features of the human mind: the ability to think ahead, plan for the future, and repress impulses.

"You see this with a lot of patients with frontal-lobe problems," Heilman says. "They can't hold back their emotions. If they get angry, then they'll just get in a fight. Even if they know that getting in a fight is a bad idea—the cognitive knowledge might still be there—that knowledge is less important than the intensity of what they are feeling." Heilman believes that in Mary's case, her damaged prefrontal cortex meant that her rational brain could no longer modulate or restrain her irrational passions. "She knew her behavior was self-destructive," Heilman says. "But she did it anyways."

The tragic story of Mary Jackson illuminates the importance of the prefrontal cortex. Because she was missing this specific brain region—it was damaged by the tumor—she couldn't think abstractly or resist her most immediate urges. She was unable to keep information in short-term memory or follow through on her long-term plans. If Mary Jackson was fleeing a fire, she never would have stopped to light the match. She would have kept on running.*

2

Imagine that you are playing a simple gambling game. You are given fifty dollars of real money and asked to decide between two options. The first option is an all-or-nothing gamble. The odds of the gamble are clear: there is a 40 percent chance that you will keep the entire fifty dollars, and a 60 percent chance that you will lose everything. The second option, however, is a sure thing. If you choose this alternative, you get to keep twenty dollars.

What option did you choose? If you're like most people, you took the guaranteed cash. It's always better to get something rather than nothing, and twenty dollars is not a trivial amount of money.

But now let's play the game again. The risky gamble hasn't changed: you still have a 40 percent chance of keeping the entire fifty dollars. This time, however, the sure thing is a loss of thirty dollars instead of a gain of twenty.

The outcome, of course, remains the same. The two gambles are identical. In both cases, you walk away with twenty of the

---

*And then there's the case of the married, middle-aged Virginia schoolteacher who suddenly started downloading child pornography and seducing young girls. His behavior was so brazen that he was quickly arrested and convicted of child molestation; he was sent to a treatment program for pedophiles, but he was expelled from the program after propositioning several women there. Having failed rehab, he was to appear in court for sentencing, but the day before his court date, he went to the emergency room complaining of blinding headaches and a constant urge to rape his neighbor. After ordering an MRI, the doctors saw the source of the problem: he had a massive tumor lodged in his frontal cortex. After the tumor was removed, the deviant sexual urges immediately disappeared. The man was no longer a hypersexual monster. Unfortunately, this reprieve was brief; the tumor started to grow back within a year. His frontal cortex was once again incapacitated, and the urges of pedophilia returned.

original fifty. But the different descriptions strongly affect how people play the game. When the choice is framed in terms of *gaining* twenty dollars, only 42 percent of people choose the risky gamble. But when the same choice is framed in terms of *losing* thirty dollars, 62 percent of people opt to roll the dice. This human foible is known as the framing effect, and it's a by-product of loss aversion, which we discussed earlier. The effect helps explain why people are much more likely to buy meat when it's labeled *85 percent lean* instead of *15 percent fat*. And why twice as many patients opt for surgery when told there's an 80 percent chance of their surviving instead of a 20 percent chance of their dying.

When neuroscientists used an fMRI machine to study the brains of people playing this gambling game, they saw the precise regions that these two different yet equivalent frames activated. They found that people who chose to gamble—the ones whose decisions were warped by the prospect of losing thirty dollars—were misled by an excited amygdala, a brain region that, when excited, evokes negative feelings. Whenever a person thinks about losing something, the amygdala is automatically activated. That's why people hate losses so much.

However, when the scientists looked at the brains of subjects who were *not* swayed by the different frames, they discovered something that surprised them. The amygdalas of these "rational" people were still active. In fact, their amygdalas tended to be just as excitable as the amygdalas of people who were susceptible to the framing effect. "We found that everyone showed emotional biases; no one was totally free of them," says Benedetto de Martino, the neuroscientist who led the experiment. Even people who instantly realized that the two different descriptions were identical—they saw through the framing effect—still experienced a surge of negative emotion when they looked at the loss frame.

What, then, caused the stark differences in behavior? If everybody had an active amygdala, why were only some people swayed by the different descriptions? This is where the prefrontal cortex enters the picture. To the surprise of the scientists, it was the activity of the prefrontal cortex (not the amygdala) that best predicted the decisions of the experimental subjects. When there was more activity in the prefrontal cortex, people were better able to resist the framing effect. They could look past their irrational feelings and realize that both descriptions were equivalent. Instead of just trusting their amygdalas, they did the arithmetic. The end result is that they made better gambling decisions. According to de Martino, "People who are more rational don't perceive emotion less, they just regulate it better."

How do we regulate our emotions? The answer is surprisingly simple: by thinking about them. The prefrontal cortex allows each of us to contemplate his or her own mind, a talent psychologists call metacognition. We know when we are angry; every emotional state comes with self-awareness attached, so that an individual can try to figure out why he's feeling what he's feeling. If the particular feeling makes no sense — if the amygdala is simply responding to a loss frame, for example — then it can be discounted. The prefrontal cortex can deliberately choose to ignore the emotional brain.

This is one of Aristotle's essential ideas. In *The Nicomachean Ethics,* his sprawling investigation into the "virtuous character," Aristotle concluded that the key to cultivating virtue was learning how to manage one's passions. Unlike his teacher Plato, Aristotle realized that rationality wasn't always in conflict with emotion. He thought Plato's binary psychology was an oversimplification. Instead, Aristotle argued that one of the critical functions of the rational soul was to make sure that emotions were intelligently applied to the real world. "Anyone can become angry — that is easy," Aristotle wrote. "But to become angry with the right person, to the right degree, at the right time, for the right purpose, and in the right way—that is not easy." That requires some thought.

One way to understand how this Aristotelian idea actually plays out in the brain is by examining the inner workings of a television focus group. Practically every show on television is tested on audiences before it hits the airwaves. When this testing process is done properly, it demonstrates a fascinating interplay between reason and emotion, instinct and analysis. In other words, the whole enterprise mimics what's constantly happening inside the human mind.

The process goes something like this: People representing a demographic cross section of America are ushered into a specially equipped room that looks like a tiny movie theater, complete with comfy seats and cup holders. (Most television focus groups take place in Orlando and Las Vegas, since those cities are full of people who have arrived from all across the country.) Each participant is given a feedback dial, a device that's about the size of a remote control and has a single red dial, a few white buttons, and a small LED screen. Feedback dials were first used in the late 1930s, when Frank Stanton, the head of audience research at CBS Radio, teamed up with Paul Lazarsfeld, the eminent sociologist, to develop the "program analyzer." The CBS method was later refined by the U.S. military during World War II as it tested its war propaganda on the public.

The modern feedback dial is designed to be as straightforward as possible so that a person can operate it without taking his or her eyes off the screen. The numbers on the dial increase in a clockwise direction, like a volume knob; higher numbers signal a more positive response to the television show. The participants are told to rotate their dials whenever their feelings change. This gives a second-by-second look at the visceral reactions of the audience, which are translated into a jagged line graph.

Although every television network depends on focus groups

for feedback—even cable channels like HBO and CNN do extensive audience research—the process has very real limits. The failures of focus groups are part of industry lore: *The Mary Tyler Moore Show, Hill Street Blues,* and *Seinfeld* are all famous examples of shows that tested terribly and yet went on to commercial success. (*Seinfeld* tested so badly that instead of being featured on NBC's 1989 fall schedule, it was introduced as a midseason replacement.) As Brian Graden, president of programming at MTV Networks, says, "Quantitative data [of the sort produced by feedback dials] is useless by itself. You've got to ask the data the right questions."

The problem with the focus group is that it's a crude instrument. People can express their feelings with dials, but they can't *explain* their feelings. The impulsive emotions recorded on the dials are just that: impulsive emotions. They are suffused with all the usual flaws of the emotional brain. Did the focus-group audience not like *Seinfeld* because they didn't like the main character? Or did they dislike the show because it was a new kind of television comedy, a sitcom about nothing in particular? (The *Seinfeld* pilot begins with a long discussion about the importance of buttons.) After all, one of the cardinal rules of focus groups is that people tend to prefer the familiar. The new shows that test the best often closely resemble shows that are already popular. For example, after the NBC sitcom *Friends* became a huge commercial hit, other networks rushed to imitate its formula. There were suddenly numerous comedy pilots about pretty twentysomethings living together in a city. "Most of these shows tested really well," one television executive told me. "The shows weren't very good, but they reminded the audience of *Friends,* which was a show they actually liked." Not one of the knockoffs was renewed for a second season.

The job of a television executive is to sort through these emotional mistakes so he or she isn't misled by the audience's first impressions. Sometimes people like shows that actually stink and reject shows that they grow to enjoy. In such situations, executives must know how to discount the responses of focus groups. They need to interpret the quantitative data, not just obey it. This is where the second-by-second responses of feedback dials are especially useful, since they allow executives to see what exactly people are responding to. A high score in minute twelve might mean that the audience really liked a particular plot twist, or it might mean that they liked looking at the blonde in her underwear. (A conclusive answer can be gotten by comparing the ratings of men versus women.) One cable channel recently tested a reality-television pilot that scored well overall but showed sharp declines in audience opinion at various points throughout the show. At first, the executives couldn't figure out what the audience didn't like. Eventually, however, they realized that the audience was reacting to the host: whenever she talked to the contestants, people turned down their dials. Although the focus-group audience said they liked the host and rated her highly when she talked to the camera, they didn't like watching her with other people. (The host was replaced.) And then there's the "flat line": when a focus-group audience is especially absorbed in the show — for example, during a climactic scene — they often forget to turn their dials. The resulting data can make it appear that the show has hit a rough spot, since many of the dials are stuck in a low position, but the reality is precisely the opposite. If the executives don't realize that the focus-group participants were simply too involved in the program to pay attention to their dials, they might end up altering the best part of the show.

The point is that the emotional data requires careful analysis. Audience research is a blunt tool, a summary of first impressions, but it can be sharpened. By examining the feelings registered on the dial, a trained observer can figure out which feelings should be trusted and which should be ignored.

This is just what the prefrontal cortex does when faced with a decision. If the emotional brain is the audience, constantly sending out visceral signals about its likes and dislikes, then the prefrontal cortex is the smart executive, patiently monitoring emotional reactions and deciding which to take seriously. It is the only brain area able to realize that the initial dislike of *Seinfeld* was a reaction to its originality, not to its inherent funniness. The rational brain can't silence emotions, but it can help figure out which ones should be followed.

IN THE EARLY 1970s, Walter Mischel invited four-year-olds to his Stanford psychology laboratory. The first question he asked each child was an easy one: did he like to eat marshmallows? The answer, not surprisingly, was always yes. Then Mischel made the child an offer. He could eat one marshmallow right away or, if the child was willing to wait for a few minutes while Mischel ran an errand, he could eat two marshmallows when the experimenter returned. Practically every child decided to wait. They all wanted more sweets.

Mischel then left the room but told the child that if he rang a bell, Mischel would come back and the child could eat the marshmallow. However, this meant that he'd be forfeiting the chance to get the second marshmallow.

Most of the four-year-olds couldn't resist the sugary temptation for more than a few minutes. Several kids covered their eyes with their hands so that they couldn't see the marshmallow. One child started kicking the desk. Another one started pulling on his hair. While a few of the four-year-olds were able to wait for up to fifteen minutes, many lasted less than one minute. Others just ate the marshmallow as soon as Mischel left the room, not even bothering to ring the bell.

The marshmallow was a test of self-control. The emotional

brain is always tempted by rewarding stimuli, such as a lump of sugar. However, if the child wanted to achieve the goal—getting a second marshmallow—then he needed to temporarily ignore his feelings, delay gratification for a few more minutes. What Mischel discovered was that even at the age of four, some kids were much better at managing their emotions than others.

Fast-forward to 1985. The four-year-olds were now high school seniors. Mischel sent out a follow-up survey to their parents. He asked the parents about a wide variety of character traits, from the ability of their child to deal with frustrating events to whether or not the child was a conscientious student. Mischel also asked for SAT scores and high school transcripts. He used this data to construct an elaborate personality profile for each of the kids.

Mischel's results were very surprising, at least to him. There was a strong correlation between the behavior of the four-year-old waiting for a marshmallow and that child's future behavior as a young adult. The children who rang the bell within a minute were much more likely to have behavioral problems later on. They got worse grades and were more likely to do drugs. They struggled in stressful situations and had short tempers. Their SAT scores were, on average, 210 points lower than those of kids who'd waited several minutes before ringing the bell. In fact, the marshmallow test turned out to be a better predictor of SAT results than the IQ tests given to the four-year-olds.

The ability to wait for a second marshmallow reveals a crucial talent of the rational brain. When Mischel looked at why some four-year-olds were able to resist ringing the bell, he found that it wasn't because they wanted the marshmallow any less. These kids also loved sweets. Instead, Mischel discovered, the patient children were better at using reason to control their impulses. They were the kids who covered their eyes, or looked in the other direction, or managed to shift their attention to something other than the delicious marshmallow sitting right there.

Rather than fixating on the sweet treat, they got up from the table and looked for something else to play with. It turned out that the same cognitive skills that allowed these kids to thwart temptation also allowed them to spend more time on their homework. In both situations, the prefrontal cortex was forced to exercise its cortical authority and inhibit the impulses that got in the way of the goal.

Studies of children with attention deficit hyperactivity disorder (ADHD) further demonstrate the connection between the prefrontal cortex and the ability to withstand emotional urges. Approximately 5 percent of school-age children are affected by ADHD, which manifests itself as an inability to focus, sit still, or delay immediate gratification. (These are the kids who eat their marshmallows right away.) As a result, kids with ADHD tend to perform significantly worse in school, since they struggle to stay on task. Minor disturbances become overwhelming distractions.

In November 2007, a team of researchers from the National Institute of Mental Health and McGill University announced that they had uncovered the specific deficits of the ADHD brain. The disorder turns out to be largely a developmental problem; often, the brains of kids with ADHD develop at a significantly slower pace than normal. This lag was most obvious in the prefrontal cortex, which meant that these kids literally lacked the mental muscles needed to resist alluring stimuli. (On average, their frontal lobes were three and a half years behind schedule.) The good news, however, is that the brain almost always recovers from its slow start. By the end of adolescence, the frontal lobes in these kids reached normal size. It's not a coincidence that their behavioral problems began to disappear at about the same time. The children who had had the developmental lag were finally able to counter their urges and compulsions. They could look at the tempting marshmallow and decide that it was better to wait.

ADHD is an example of a problem in the developmental process, but the process itself is the same for everybody. The maturation of the human mind recapitulates its evolution, so the first parts of the brain to evolve — the motor cortex and brain stem — are also the first parts to mature in children. Those areas are fully functional by the time humans hit puberty. In contrast, brain areas that are relatively recent biological inventions — such as the frontal lobes — don't finish growing until the teenage years are over. The prefrontal cortex is the last brain area to fully mature.

This developmental process holds the key to understanding the behavior of adolescents, who are much more likely than adults to engage in risky, impulsive behavior. More than 50 percent of U.S. high school students have experimented with illicit drugs. Half of all reported cases of sexually transmitted diseases occur in teenagers. Car accidents are the leading cause of death for those under the age of twenty-one. These bleak statistics are symptoms of minds that can't restrain themselves. While the emotional brains of teens are operating at full throttle (those raging hormones don't help), the mental muscles that check these emotions are still being built. A recent study by neuroscientists at Cornell, for example, demonstrated that the nucleus accumbens, a brain area associated with the processing of rewards — things like sex, drugs, and rock 'n' roll — was significantly more active and mature in the adolescent brain than the prefrontal cortex was, that part of the brain that helps resist such temptations. Teens make bad decisions because they are literally less rational.*

---

*But there are ways to compensate for the irrational brains of teens. For instance, when West Virginia revoked driving permits for students who were under the age of eighteen and who dropped out of school, the dropout rate fell by one-third in the first year. While teens were blind to the long-term benefits of getting a high school diploma, they could appreciate the short-term punishment of losing a license. The New York City schools have recently begun experimenting with a program that pays students for improving their standardized test scores; initial results have been extremely encouraging. By focusing on immediate rewards, these incentive programs help correct for the immature prefrontal cortices of children and teenagers.

This new research on reckless adolescents and children with ADHD highlights the unique role of the prefrontal cortex. For too long, we've assumed that the purpose of reason is to eliminate those emotions that lead us astray. We've aspired to the Platonic model of rationality, in which the driver has complete control. But now we know that silencing human feelings isn't possible, at least not directly. Every teenager wants to have sex, and every four-year-old wants to eat marshmallows. Every firefighter who sees a wall of flames wants to run. Human emotions are built into the brain at a very basic level. They tend to ignore instructions.

But this doesn't mean that humans are mere puppets of the limbic system. Some people can see through the framing effect despite the fact that their amygdalas are activated. Some four-year-olds can find ways to wait for the second marshmallow. Thanks to the prefrontal cortex, we can transcend our impulses and figure out which feelings are useful and which ones should be ignored.

Consider the Stroop task, one of the classic experiments of twentieth-century psychology. Three words—*blue, green,* and *red*—are flashed randomly on a computer screen. Each of the words is printed in a different color, but the colors aren't consistent. The word *red* might be in green, while *blue* is in red. The surprisingly difficult job of the subject is to ignore the *meaning* of the word and focus instead on the *color* of the word. If you're looking at *green,* but the word is actually in blue letters, then you have to touch the button marked *blue.*

Why is this simple exercise so hard? Reading the word is an automated task; it takes little mental effort. Naming the color of the word, however, requires deliberate thought. The brain needs to turn off its automatic operation—the act of reading a familiar word—and consciously think about what color it sees. When a person performs the Stroop task in an fMRI machine, scientists can watch the brain struggle to ignore the obvious answer. The most important cortical area engaged in this tug of war is the prefrontal cortex, which allows a person to reject the first impression when it's possible that the first impression might be wrong. If the emotional brain is pointing you in the direction of a bad decision, you can choose to rely on your rational brain instead. You can use your prefrontal cortex to discount the amygdala, which is telling you to run up the steep slopes of the gulch. The reason Wag Dodge survived was not that he wasn't scared. Like all the smokejumpers, he was terrified. Dodge survived because he realized that his fright wasn't going to save him.

## 3

The ability to supervise itself, to exercise authority over its own decision-making process, is one of the most mysterious talents of the human brain. Such a mental maneuver is known as executive control, since thoughts are directed from the top down, like a CEO issuing orders. As the Stroop task demonstrates, this thought process depends on the prefrontal cortex.

But the questions still remain: How does the prefrontal cortex wield such power? What allows this particular area to control the rest of the brain? The answer returns us to the cellular details: by looking at the precise architecture of the prefrontal cortex, we can see the neural forms that explain its function.

Earl Miller is a neuroscientist at MIT who has devoted his career to understanding this bit of tissue. He was first drawn to the prefrontal cortex as a graduate student, in large part because it seemed to be connected to *everything*. "No other brain area gets so many different inputs or has so many different outputs," Miller says. "You name the brain area, and the prefrontal cortex is almost certainly linked to it." It took more than a decade of painstaking probing while Miller carefully monitored cells all across the monkey brain, but he was eventually able to show that the prefrontal cortex wasn't simply an aggregator of information. Instead, it was like the conductor of an orchestra, waving its baton and directing the musicians. In 2007, in a paper published in *Science,* Miller was able to provide the first glimpse of executive control at the level of individual neurons, as cells in the prefrontal cortex directly modulated the activity of cells throughout the brain. He was watching the conductor at work.

However, the prefrontal cortex isn't merely the bandleader of the brain, issuing one command after another. It's also uniquely versatile. While every other cortical region is precisely tuned for specific kinds of stimuli—the visual cortex, for example, can deal only with visual information—the cells of the prefrontal cortex are extremely flexible. They can process whatever kind of data they're told to process. If someone is thinking about an unfamiliar math problem on a standardized test, then her prefrontal neurons are thinking about that problem. And when her attention shifts, and she starts to contemplate the next question on the test, these task-dependent cells seamlessly adjust their focus. The end result is that the prefrontal cortex lets her consciously analyze any type of problem from every possible angle. Instead of responding to the most obvious facts, or the facts that her emotions think are most important, she can concentrate on the facts that might help her come up with the right answer. We can all use executive control to get creative, to think about the same old problem in a new way. For instance, once Wag Dodge realized that he couldn't outrun the flames and that the fire would beat the smokejumpers to the top of the ridge, he needed to use his prefrontal cortex to come up with a new solution. The obvious response wasn't going to work. As Miller notes, "That Dodge guy had some high prefrontal function."

Consider the classic psychology puzzle known as the "candle problem." A subject is given a book of matches, some candles, and a cardboard box containing a few thumbtacks. The person is told to attach the candle to a piece of corkboard in such a way that it can burn properly. Most people initially attempt two common strategies, neither of which will work. The first strategy is to tack the candle directly to the board; this causes the candle wax to shatter. The next is to use the matches to melt the bottom of the candle and then try to stick the candle to the board; the wax does not hold, and the candle falls to the floor. At this point, most people give up. They tell the scientists that the puzzle is impossible; it's a stupid experiment and a waste of time. Less than 20 percent of people manage to come up with the correct solution, which is to attach the candle to the cardboard box and then tack the cardboard box to the board. Unless the subject has an insight about the box—that it can do more than hold thumbtacks—candle after candle will be wasted. The subject repeats his failures while waiting for a breakthrough.

People with frontal-lobe lesions can never solve puzzles like the candle problem. Although they understand the rules of the game, they are completely unable to think creatively about the puzzle, to look past their initial (and incorrect) answers. The end result is that the frontal-lobe patient fails to execute the counterintuitive moves required to solve the puzzle, even though the obvious moves have failed. Instead of trying something new, or relying on abstract thought, the subject keeps attempting to tack the candle to the board, stubbornly insisting on this strategy until there are no more candles.

Mark Jung-Beeman, a cognitive psychologist at Northwestern University, has spent the last fifteen years trying to understand how the brain, led by the prefrontal cortex, manages to come up with such creative solutions. He wants to find the neural source of our breakthroughs. Jung-Beeman's experiments go like this: he gives a subject three different words (such as *pine, crab,* and *sauce*) and asks him to think of a single word that could form a compound word or phrase with all three. (In this case, the answer is *apple: pineapple, crab apple, applesauce.*) What's interesting about this type of verbal puzzle is that the answers often arrive in a flash of insight, the familiar "aha!" moment. People have no idea how they came up with the necessary word, just as Wag Dodge couldn't explain how he invented the escape fire. Nevertheless, Jung-Beeman found that the mind was carefully preparing itself for the epiphany; every successful insight was preceded by the same sequence of cortical events. (He likes to quote Louis Pasteur: "Chance favors the prepared mind.")

The first brain areas activated during the problem-solving process were those involved with executive control, such as the prefrontal cortex and anterior cingulate cortex. The brain was banishing irrelevant thoughts so that the task-dependent cells could properly focus. "You're getting rid of those errant daydreams and trying to forget about the last word puzzle you worked on," Jung-Beeman says. "Insight requires a clean slate."

After exercising top-down control, the brain began generating associations. It selectively activated the necessary brain areas, looking for insights in all the relevant places, searching for the association that would give the answer. Because Jung-Beeman was giving people word puzzles, he saw additional activation in areas related to speech and language, such as the superior temporal gyrus in the right hemisphere. (The right hemisphere is particularly good at generating the kind of creative associations that lead to epiphanies.) "Most of the possibilities your brain comes up with aren't going to be useful," he says. "And it's up to the executive-control areas to keep on looking or, if necessary, change strategies and start looking somewhere else."

But then, when the right answer suddenly appeared—when *apple* was passed along to the frontal lobes—there was an immediate realization that the puzzle had been solved. "One of the interesting things about such moments of insight," says Jung-Beeman, "is that as soon as people have the insight, they say it just seems obviously correct. They know instantly that they've solved the problem."

This act of recognition is performed by the prefrontal cortex, which lights up when a person is shown the right answer, even if he hasn't come up with the answer himself. Of course, once the insight has been identified, those task-dependent cells in the frontal lobes immediately move on to the next task. The mental slate is once again wiped clean. The brain begins preparing itself for another breakthrough.

ON THE AFTERNOON of July 19, 1989, United Airlines Flight 232 took off from Denver Stapleton Airport, bound for Chicago. The conditions for the flight were ideal. The morning thunderstorms had passed, and the sky was a cloudless cerulean blue. Once the DC-10 reached its cruising altitude of 37,000 feet, about thirty minutes after takeoff, Captain Al Haynes turned off the seat-belt sign. He didn't expect to turn it back on until the plane began its descent.

The first leg of the flight went smoothly. A hot lunch was served to the passengers. The plane was put on autopilot, with supervision by the first officer, William Records. Captain Haynes drank a cup of coffee and stared at the cornfields of Iowa far below. He'd flown this exact route dozens of times before — Haynes was one of United's most experienced pilots, with more than thirty thousand hours of flight time — but he never ceased to admire the grid of flat land, the farms laid out in such perfectly straight lines.

At 3:16 in the afternoon, about an hour after takeoff, the quiet of the cockpit was shattered by the sound of a loud explosion coming from the back of the plane. The frame of the aircraft shuddered and lurched to the right. Haynes's first thought was that the plane was breaking up, that he was about to die in a massive fireball. But then, after a few seconds of gnashing metal, the quiet returned. The plane kept on flying.

Haynes and First Officer Records immediately began scanning the cluster of instruments and dials, looking for some indication of what had gone wrong. The pilots noticed that the number two engine, the middle engine in the rear of the plane, was no longer operating. (Such a failure can be dangerous, but it's rarely catastrophic, since the DC-10 also has two other engines, one on each wing.) Haynes got out his pilot manual and started going through the engine-failure checklist. The first order of business was to shut off the fuel supply to that engine, in order to minimize the risk of an engine fire. They attempted it, but the fuel lever wouldn't move.

It had now been a few minutes since the explosion. Records was flying the plane. Haynes was still trying to fix the fuel lines; he assumed that the plane was maintaining its scheduled flight path to Chicago, albeit at a slightly slower pace. That's when Records turned to him and said the one thing a pilot never wants to hear: "Al, I can't control the airplane." Haynes looked over at Records, who had applied full left aileron and pushed the yoke so far forward that the controls were pressed against the cockpit dash. Under normal circumstances, such a maneuver would have caused the plane to descend and turn left. Instead, the plane was in a steep ascent with a sharp right bank. If the plane banked much more, it would flip over.

What could trigger such a complete loss of control? Haynes assumed there had been a massive electronic failure, but the circuit board looked normal. So did the onboard computers. Then Haynes checked the pressure on his three hydraulic lines: they were all plummeting toward zero. "I saw that and my heart skipped a beat," Haynes remembers. "It was an awful moment, the first time I realized that this was a real disaster." The hydraulic systems control the plane. They are used to adjust everything from the rudder to the wing flaps. Planes are always engineered with multiple, fully independent hydraulic systems; if one fails, the backup system can take its place. This redundancy means that a catastrophic failure of all three lines simultaneously should be virtually impossible. Engineers calculate the odds of such an event at about a billion to one. "It wasn't something we ever trained for or practiced," Haynes says. "I looked in my pilot manual, but there was nothing about a total loss of hydraulics. It just wasn't supposed to happen."

But that's exactly what had happened to this DC-10. For some reason, the loss of the engine had ruptured all three hydraulic lines. (Investigators later discovered that the engine fan disc had fractured, sending shards of metal through the tail section where all the hydraulic lines were located.) Haynes could remember only one other instance when an aircraft had lost all of its hydraulic controls. Japan Airlines Flight 123, a Boeing 747 flying from Tokyo to Osaka in August 1985, had suffered a similar catastrophe after its vertical stabilizer was blown off by an explosive decompression event. The aircraft had steadily drifted downward for more than thirty minutes, eventually crashing into the face of a mountain. More than five hundred people died. It was the deadliest single-aircraft disaster in history.

Back in the cabin, the passengers were beginning to panic. Everyone had heard the explosion; they all could feel the plane careening out of control. Dennis Fitch, a United Airlines flight instructor, was sitting in the middle of the aircraft. After the terrifying boom — "It sounded like the plane was breaking apart," Fitch said — he visually inspected the wings of the plane. There were no obvious signs of damage, although he couldn't figure out why the pilots weren't correcting the plane's steep bank. Fitch knocked on the cockpit doors to see if he could offer any assistance. He taught pilots how to fly the DC-10, so he knew the aircraft inside and out.

"It was an amazing scene," Fitch remembers. "Both pilots were at the controls, their tendons in their forearms were raised from effort, their knuckles were white from gripping the handles, but it wasn't doing anything." When the pilots told Fitch that they had lost hydraulic pressure in all three hydraulic systems, Fitch was shocked. "There was no procedure for this. When I heard that, I thought, *I'm going to die this afternoon.*"

Captain Haynes, meanwhile, was desperately trying to think of some way to regain control. He placed a radio call to United Airlines' System Aircraft Management (SAM), a crew of aircraft engineers specially trained to help deal with in-flight emergencies. "I thought, these guys must know a way out of this mess," Haynes says. "That's their job, right?"

But the engineers at SAM weren't any help. For starters, they didn't believe that all of the hydraulic pressure was really gone. "SAM kept on asking us to check the hydraulics again," Haynes says. "They told us that there must be some pressure left. But I kept on telling them that there was none. All three lines were empty. And then they kept on telling us to check the pilot's manual, but the manual didn't deal with this problem. Eventually, I realized that we were on our own. Nobody was going to land the plane for us."

Haynes began by making a mental list of the cockpit elements that he could operate without hydraulic pressure. The list was short. In fact, Haynes could think of only one element that might still be useful: the thrust levers, which controlled the speed and power of his two remaining engines. (They are like the gas pedals of the plane.) But what does thrust matter if you can't maneuver? It would be like revving a car without a steering wheel.

Then Haynes had an idea. At first, he dismissed it as crazy. The more he thought about it, however, the less ridiculous it seemed. His idea was to use his thrust levers to steer the plane. The key was differential thrust; *thrust* is the forward-directed force of an airplane engine, and a difference in thrust between the plane's engines is normally something pilots want to avoid. But Haynes figured that if he idled one engine while the other got a boost of power, the plane should turn to the idled side. The idea was grounded in simple physics, but he had no idea if it would actually work.

There was little time to lose. The bank of the plane was approaching 38 degrees. If it got past 45 degrees, the plane would flip over and enter a death spiral. So Haynes advanced the throttle for the right engine and idled the left. At first, nothing happened. The plane stayed in a steep bank. But then, ever so slowly, the right wing began to level itself. The plane was now flying in a straight line. Haynes's desperate idea had worked.

Flight 232 was given instructions to land at Sioux City, Iowa, a regional airport about ninety miles to the west. Using nothing but variations in engine thrust, the pilots began a steady right-hand turn. It had been about twenty minutes since the initial explosion, and it seemed as if Haynes and his crew had restored a measure of control to the uncontrollable plane. "I felt like we were finally making some progress," Haynes says. "It was the first time since the explosion that I thought we just might be able to get this bird on the ground."

But just as the flight crew was starting to gain a little confidence, the plane started to pitch violently up and down in a relentless cycle. This is known as a phugoid pattern. Under normal flight conditions, phugoids are easy to manage, but since the plane was without any hydraulic pressure, Haynes and his crew were unable to modulate the pitch of the aircraft. The pilots realized that unless they found a way to dampen the phugoids, they could end up like the Japan Airlines' Flight 123. They would careen in a sine wave as they steadily lost altitude. And then they'd crash into the cornfields.

How do you control phugoids in such a situation? At first glance, the answer seems obvious. When the nose of the plane is pitched down, and the air speed is increasing, a pilot should decrease the throttle, so that the plane slows down. And when the plane is pitched up, and the air speed is decreasing, a pilot should increase the throttle in order to prevent a stall. "You're looking at your air-speed indicator, and the natural reaction of a pilot is to try to balance out what's happening," Haynes says. But that instinctive reaction is exactly the opposite of what should be done. The aerodynamics of flight contradict common sense; if Haynes had gone with his first impulse, he would soon have lost control of the plane. The aircraft would have entered a steep, unstoppable descent.

Instead of doing that, Haynes carefully thought through the problem. "I tried to imagine what would happen to the plane depending on how I controlled the thrust levers," he says. "It took me a few moments, but that saved me from making a big mistake." Haynes realized that when the nose tilted down and the air speed built up, he needed to *increase* power, so that the two remaining engines could bring up the nose. Because the engines on a DC-10 are set below the wings, an increase in engine throttle will cause the plane to pitch up. In other words, he needed to accelerate on the downhill and brake on the uphill. It was such a counterintuitive idea that Hayes could barely bring himself to execute the plan. "The hardest part," Haynes said, "was when the nose started up and the air speed started to fall, and then you had to close the throttles. That wasn't very easy to do. You felt like you were going to fall out of the sky."

But it worked. The pilots managed to keep the plane reasonably level. They couldn't get rid of the phugoid motion—that would have required actual flight controls—but they kept it from turning into a deadly dive. The flight crew was now able to focus on their final problem: orchestrating a descent into Sioux City. Haynes knew it would be a struggle. For one thing, the pilots couldn't directly control their rate of descent, since the elevators of the aircraft—the control surfaces in the tail wing of the plane that modulate altitude—were completely unresponsive. Haynes and the pilots were forced to rely on a rough formula used when flying the DC-10: a thousand-foot drop in altitude takes approximately three miles in distance. Since the aircraft was now about sixty miles from the airport but was maintaining an altitude of approximately thirty thousand feet, Haynes realized they'd need to make a few loops on their way to the runway. If they tried to rush the descent, they'd risk losing what little stability they had. And so the pilots began a series of right-hand turns as they proceeded northwest to Sioux City. With each turn, they lost a little more altitude.

As the plane neared the airport, the pilots made final preparations for an emergency landing. Excess fuel was dumped and the throttles were gradually eased. The passengers were told to assume the brace position, with their heads tight against their knees. Haynes could see the landing strip and the fire engines in the distance. Although the pilots had been flying without controls for forty minutes, they still managed to line up the plane in the middle of the runway, with its wheels down and its nose up. It was an incredible feat of airmanship.

Unfortunately, the pilots had no control over the speed of the plane. They also couldn't brake once they hit the runway. "You normally land the DC-10 at approximately a hundred and forty knots," Haynes says. "We were doing two hundred and fifteen knots and accelerating. You normally touch down at about two to three hundred feet per minute at the most, as a rate of descent. We were doing eighteen hundred and fifty feet per minute. And increasing. And you normally like to go straight down the runway, and we were drifting left and right because of the tail wind."

These factors meant that the plane couldn't stay on the tarmac. It skidded through a cornfield and shattered into several sections. The cockpit broke apart from the main body of the plane, like the tip of a pencil, and tumbled end over end to the edge of the airfield. (All of the pilots were knocked unconscious and suffered life-threatening injuries.) A fire broke out in the fuselage. Toxic black smoke filled the main cabin. When the smoke cleared, 112 passengers were dead.

But the piloting skills of the flight crew—their ability to control a plane without any controls—meant that 184 passengers survived the accident. Because the plane made it to the airport, emergency responders were able to treat the wounded and quickly extinguish the flames. As the National Transportation Safety Board concluded in their authoritative report, "The performance [of the pilots] was highly commendable and greatly exceeded reasonable expectations." The method of flight control invented in the cockpit of Flight 232 is now a standard part of pilot training.

4

The first remarkable thing about the performance of the pilots is that they managed to keep their emotions in check. It's not easy to maintain poise when you've lost complete control of your aircraft. In fact, Haynes later admitted that he didn't expect to survive the flight. He assumed that Flight 232 would eventually spiral out of control, that the phugoids would get worse and worse until the plane finally crashed into the ground. "I thought the best-case scenario was that we'd make the runway but crashland," Haynes says. "And I was still pretty sure that I wouldn't survive that."

And yet, Haynes never let his fear turn into panic. He was in a situation of incomprehensible pressure, confronted with a scenario that was never supposed to happen, but he managed to keep his cool. Such restraint was possible only because Haynes, like Wag Dodge, used his prefrontal cortex to manage his emotions. After the three hydraulic lines failed, the pilot realized that his trained instincts didn't know how to land the plane. Emotions are adept at finding patterns based on experience, so that a person can detect the missile amid the blur of radar blips. But when you encounter a problem you've never experienced before, when your dopamine neurons have no idea what to do, it's essential that you try to tune out your feelings. Pilots call such a state "deliberate calm," because staying calm in high-pressure situations requires conscious effort. "Maintaining our composure was one of the hardest things we had to do," Haynes says. "We knew we had to focus and think straight, but that's not always so easy."

Preventing the onset of panic, however, was only the first step. If Haynes and his crew were going to land the plane at Sioux City, they needed to improvise a solution to their unprecedented problem. Consider the use of differential thrust. Such a method of flight control had never been attempted before. Haynes had never practiced it in a simulator or even contemplated the possibility of turning using only his engines. Even the SAM engineers didn't know what to do. And yet, in the terrifying moments after the explosion, when Haynes looked at his dash and saw that he had no central engine and no hydraulic pressure, he was able to figure out a way to keep the plane in the air.

It's worth taking a closer look at this single decision so that we can better understand what, exactly, allows the prefrontal cortex to deal with such fraught situations. Steven Predmore, a manager of human-factors analysis at Delta Airlines, has studied the decision-making process during Flight 232 in exquisite detail. He began by breaking down the thirty-four minutes of conversation captured by the cockpit voice recorder into a series of thought units, or pieces of information. By analyzing the flow of these thought units, Predmore was able to map out the sequence of events from the perspective of the pilots.

Predmore's study is a gripping portrait of heroism and teamwork. Shortly after Haynes realized that the plane had lost all hydraulic pressure, the air-traffic controllers began consulting with the pilots on the best flight path into Sioux City. Haynes's advice was simple: "Whatever you do," he said, "keep us away from the city." At other moments, the transcripts reveal the pilots struggling to lighten the mood:

FITCH: I'll tell you what, we'll have a beer when this is done.
HAYNES: Well, I don't drink, but I'll sure as hell have one.

And yet, even as the pilots were cracking jokes, they were making difficult decisions under extreme cognitive stress. During the descent into Sioux City, the number of thought units exchanged in the cockpit consistently exceeded thirty per minute, with peaks of nearly sixty per minute. That's nearly one new piece of information every second. (Under normal flight conditions, the number of thought units rarely exceeds ten per minute.) Some of this information was critical—the pilots closely followed their altitude levels—and some of it was less relevant. After all, it doesn't really matter how the yoke is positioned if the yoke is broken.

The pilots dealt with this potential information overload by quickly focusing on the most necessary bits of data. They were always thinking about what they should think about, which let them minimize potential distractions. For instance, once Haynes realized that he could control only the throttle levers—everything else in the cockpit was virtually useless—he immediately zeroed in on the possibility of steering with his engines. He stopped worrying about his ailerons, elevators, and wing flaps. Once the plane was within twenty miles of the Sioux City airport, about twelve minutes from touchdown, the captain started to concentrate on executing the landing. He deliberately ignored everything else. According to Predmore, the ability of the flight crew to successfully prioritize their tasks was a crucial ingredient of their success.

Of course, it's not enough to just think about a problem;

Haynes needed to *solve* his problem, to invent a completely new method of flight control. This is where the prefrontal cortex really demonstrates its unique strengths. It is the only brain region able to take an abstract principle—in this case, the physics of engine thrust—and apply it in an unfamiliar context to come up with an entirely original solution. It's what allowed Haynes to logically analyze the situation, to imagine his engines straightening his steep bank. He could model the aerodynamics in his mind.

Only recently have scientists learned how the prefrontal cortex accomplishes this. The key element is a special kind of memory known as working memory. The name is accurate: by keeping information in short-term storage, where it can be manipulated and analyzed, the brain can work with all the information streaming in from other cortical areas. It is able to determine what information, if any, is relevant to the problem it's trying to solve. For instance, studies show that neurons in the prefrontal areas will fire in response to a stimulus—such as the sight of some cockpit instrumentation—and then keep on firing for several seconds after the stimulus has disappeared. This echo of activity allows the brain to make creative associations as seemingly unrelated sensations and ideas overlap. (Scientists refer to this as the restructuring phase of problem-solving, since the relevant information is mixed together in new ways.) It's why Haynes could think about the thrust levers while simultaneously thinking about how to turn the plane. Once this overlapping of ideas occurs, cortical cells start to form connections that have never existed before, wiring themselves into entirely new networks. And then, after the insight has been generated, the prefrontal cortex is able to identify it: you immediately realize that this is the answer you've been searching for. "I don't know where the idea for differential thrust came from," Haynes says. "It just occurred to me, all of a sudden, out of nowhere." From the perspective of the brain, new ideas are merely several old thoughts that occur at the exact same time.

The problem-solving abilities of working memory and the prefrontal cortex are a crucial aspect of human intelligence. Numerous studies have found strong correlations between scores on tests of working memory and tests of general intelligence. Being able to hold more information in the prefrontal cortex, and being able to hold on to that information for longer, means that brain cells are better able to form useful associations. At the same time, the rational brain must also stringently filter out all extraneous thoughts, since they might lead to unhelpful connections. Unless you are disciplined about what you choose to think about—and the pilots of Flight 232 were extremely disciplined—you won't be able to effectively think through your problem. You'll be so overwhelmed by all those incoming ideas that you'll never be able to figure out which ones are genuine insights.

Look, for example, at the phugoids. When the aircraft started to pitch up and down, Haynes's first impulse was to increase the throttle when the plane was ascending, so that the plane maintained air speed. But then Haynes made himself think, for a few extra seconds, about the implications of this approach. He blocked out all the other things he could have been worrying about—he still didn't know how he was going to land the plane—and focused instead on the relationship of his thrust levers and the pitch of the plane. That's when Haynes realized that trusting his instincts in this situation was a deadly mistake. His explicit analysis, made possible by working memory, allowed him to come up with a new solution. If the plane was going up, then he needed to slow down.

Such decision-making is the essence of rationality. In the months after Flight 232, the United training center in Denver commissioned numerous pilots, including a test pilot from Mc-

Donnell Douglas, to see if anyone could land a DC-10 without hydraulics. The training center used a flight simulator that was programmed with the precise conditions faced by the United crew on that July day. "These other pilots kept trying to land the plane at Sioux City, just like we did," says Haynes. "But they always had some kind of unfortunate event and kept on crashing outside the airport." In fact, the pilots trying to land the DC-10 in the simulator failed to make the runway on their first *fifty-seven* attempts.

Haynes is a modest man; he says most of the passengers survived because of "luck and teamwork." However, the landing of Flight 232 on the Sioux City runway was clearly a case of Haynes creating his own luck. Because he took advantage of his prefrontal cortex, relying on its uniquely flexible neurons, he managed to avert an almost certain disaster. He was able to maintain his cool and analyze the situation in a deliberate manner so that he could generate the necessary flash of insight. "I'm no genius," Haynes says. "But a crisis like that sure can sharpen the mind."

Although the rational talents of the prefrontal cortex kept Flight 232 from crashing into a cornfield, it's important to realize that rationality isn't an all-purpose solution. In the next chapter, we are going to look at what happens when people use their prefrontal cortices in the wrong way. It's possible to think too much.

# 5

# Choking on Thought

The lesson of Wag Dodge, television focus groups, and Flight 232 is that a little rational thought can save the day. In such situations, the prefrontal cortex is uniquely designed to come up with creative answers, to generate that flash of insight that leads a person to the right decision. Such narratives fit comfortably with our broad assumption that more deliberation is always better. In general, we believe that carefully studying something leads to better outcomes, since we'll avoid careless errors. Consumers should always comparison shop so that they find the best products. Before we invest in stocks, we are supposed to learn as much as possible about the company. We expect doctors to order numerous diagnostic tests, even if the tests are expensive and invasive. In other words, people believe that a decision that's the result of rational deliberation will always be better than an impulsive decision. This is why one shouldn't judge a book by its cover or propose marriage on the first date. When in doubt, we try to resort to careful analysis and engage the rational circuits of the prefrontal cortex.

This faith in the power of reason is easy to understand. Ever since Plato, we've been assured that a perfectly rational world would be a perfect world, a Shangri-la ruled by statistical equations and empirical evidence. People wouldn't run up credit card debt or take out subprime loans. There would be no biases or prejudices, just cold, hard facts. This is the utopia dreamed of by philosophers and economists.

However, this new science of decision-making (a science rooted in the material details of the brain) is most interesting when the data turns out to contradict the conventional wisdom. Ancient assumptions are revealed as just that: assumptions. Untested theories. Unsubstantiated speculation. Plato, after all, didn't do experiments. He had no way of knowing that the rational brain couldn't solve every problem, or that the prefrontal cortex had severe limitations. The reality of the brain is that, sometimes, rationality can lead us astray.

FOR RENEE FLEMING, the opera superstar, the first sign of trouble came during a routine performance of Mozart's *The Marriage of Figaro* at the Lyric Opera of Chicago. Fleming was singing the "Dove sono" aria from act 3, one of the most beloved songs in all of opera. At first, Fleming sang Mozart's plaintive melody with her typical perfection. She made the high notes sound effortless, her voice capturing the intensity of emotion while maintaining her near perfect pitch. Most sopranos struggle with Mozart's tendency to compose in the *passaggio*, or the awkward part of the vocal range between registers. But not Fleming. Her performance the night before had earned her a long standing ovation.

But then, just as she neared the most difficult section of the aria—a crescendo of fluttering pitches, in which her voice has to echo the violins—Fleming felt a sudden stab of self-doubt. She couldn't stop thinking that she was about to make a mistake. "It caught me by surprise," she later wrote in her memoir. "That aria was never an easy piece, but it was certainly one with which I had had an enormous amount of experience." In fact, Fleming had performed this piece hundreds of times before. Her first big operatic break had been singing the role of the Countess at the Houston Opera, more than a decade earlier. The tragic "Dove sono" aria, in which the Countess questions the loss of her happiness, had been featured on Fleming's first album and became a standard part of her repertoire. It was, Fleming said, her "signature piece."

And yet now, she could barely breathe. She felt her diaphragm constrict, sucking the power from her voice. Her throat tightened and her pulse started to race. Although Fleming fought her way through the rest of the song, stealing breaths wherever possible—she still managed to get a standing ovation—she was deeply shaken. What had happened to her self-confidence? Why did her favorite aria suddenly make her so nervous?

Before long, Fleming's performance problems became chronic. The songs that used to be second nature were suddenly impossible to sing. Every performance was a struggle against anxiety, against that monologue in her head telling her not to make a mistake. "I had been undermined by a very negative inner voice," she wrote, "a little nattering in my ear that said, 'Don't do that . . . Don't do this . . . Your breath is tight . . . Your tongue has gone back . . . Your palate is down . . . The top is spread . . . Relax your shoulders!'" Eventually, it got so bad that Fleming planned to quit opera altogether. She was one of the most talented performers in the world, and yet she could no longer perform.

Performers call such failures "choking," because a person so frayed by pressure might as well not have any oxygen. What makes choking so morbidly fascinating is that the only thing incapacitating the performer is his or her own thoughts. Fleming, for example, was so worried about hitting the high notes of Mozart's opera that she failed to hit them. The inner debate over proper technique made her voice seize up, and it became impossible to sing with the necessary speed and virtuosity. Her mind was sabotaging itself.

What causes choking? Although it might seem like an amorphous category of failure, or even a case of excess emotion, choking is actually triggered by a specific mental mistake: *thinking too much*. The sequence of events typically goes like this: When a person gets nervous about performing, he naturally becomes extra self-conscious. He starts to focus on himself, trying to make sure that he doesn't make any mistakes. He begins scrutinizing actions that are best performed on autopilot. Fleming started to think about aspects of singing that she hadn't thought about since she was a beginner, such as where to position her tongue and how to shape her mouth for different pitches. This kind of deliberation can be lethal for a performer. The opera singer forgets how to sing. The pitcher concentrates too much on his motion and loses control of his fastball. The actor gets anxious about his lines and seizes up onstage. In each of these instances, the natural fluidity of performance is lost. The grace of talent disappears.

Consider one of the most famous chokes in sports history: the collapse of Jean Van de Velde on the last hole of the 1999 British Open. Until that point in the tournament, Van de Velde had been playing nearly flawless golf. He had a three-stroke lead entering the eighteenth hole, which meant that he could double-bogey (that is, be two strokes over par) and still win. On his previous two rounds, he'd birdied (been one stroke under par) this very hole.

Now Van de Velde was the only player on the course. He knew that the next few shots could change his life forever, turning a PGA journeyman into an elite golfer. All he had to do was play it safe. During his warm-up swings on the eighteenth, Van de Velde looked nervous. It was a blustery Scotland day, but beads of sweat were glistening on his face. After repeatedly wiping away the perspiration, he stepped up to the tee, planted his feet, and jerked back his club. His swing looked awkward. His hips spun out ahead of his body, so that the face of his driver wasn't straight on the ball. Van de Velde watched the white speck sail away and then bowed his head. He had bent the ball badly to the right, and it ended up twenty yards from the fairway, buried in the rough. On his next shot, he made the same mistake, but this time he sent the ball so far right that it bounced off the grandstands and ended up in a patch of knee-high grass. His third shot was even worse. By this point, his swing was so out of sync that he almost missed the ball; it was launched into the air along with a thick patch of grass. As a result, his shot came up far short and plunged into the water hazard just before the green. Van de Velde grimaced and turned away, as if he couldn't bear to watch his own collapse. After taking a penalty, he was still sixty yards short of the hole. Once again, his tentative swing was too weak, and the ball ended up exactly where he didn't want it: in a sandy bunker. From there, he managed to chip onto the green and, after *seven* errant shots, finish the round. But it was too late. Van de Velde had lost the British Open.

The pressure of the eighteenth hole was Van de Velde's undoing. When he started thinking about the details of his swing, his swing broke down. On the last seven shots, Van de Velde seemed like a different golfer. He had lost his easy confidence. Instead of playing like a pro on the PGA tour, he started swinging with the cautious deliberation of a beginner with a big handicap. He was suddenly focusing on the mechanics of his stroke, making sure that he didn't torque his wrist or open his hips. He was literally regressing before the crowd, reverting to a mode of explicit thought that he hadn't used on the golf green since he was a child learning how to swing.

Sian Beilock, a professor of psychology at the University of Chicago, has helped illuminate the anatomy of choking. She uses putting on the golf green as her experimental paradigm. When people are first learning how to putt, the activity can seem daunting. There are just so many things to think about. A golfer needs to assess the lay of the green, calculate the line of the ball, and get a feel for the grain of the turf. Then the player has to monitor the putting motion and make sure the ball is hit with a smooth, straight stroke. For an inexperienced player, a golf putt can seem impossibly hard, like a life-size trigonometry problem.

But the mental exertion pays off, at least at first. Beilock has shown that novice putters hit better shots when they consciously reflect on their actions. The more time the beginner spends thinking about the putt, the more likely he is to sink the ball in the hole. By concentrating on the golf game, by paying attention to the mechanics of the stroke, the novice can avoid beginners' mistakes.

A little experience, however, changes everything. After a golfer has learned how to putt—once he or she has memorized the necessary movements—analyzing the stroke is a waste of time. The brain already knows what to do. It automatically computes the slope of the green, settles on the best putting angle, and decides how hard to hit the ball. In fact, Beilock found that when experienced golfers are forced to think about their putts, they hit significantly *worse* shots. "We bring expert golfers into our lab, and we tell them to pay attention to a particular part of their swing, and they just screw up," Beilock says. "When you are at a high level, your skills become somewhat automated. You don't need to pay attention to every step in what you're doing."

Beilock believes that this is what happens when people "choke." The part of the brain that monitors behavior—a network centered in the prefrontal cortex—starts to interfere with decisions that are normally made without thinking. It begins second-guessing the skills that have been honed through years of diligent practice. The worst part about choking is that it tends to be a downward spiral. The failures build on one another, and a stressful situation is made even more stressful. After Van de Velde lost the British Open, his career hit the skids. Since 1999, he has failed to finish in the top ten in a major tournament.*

Choking is merely a vivid example of the havoc that can be caused by too much thought. It's an illustration of rationality gone awry, of what happens when we rely on the wrong brain areas. For opera singers and golf players, such deliberate thought processes interfere with the trained movements of their muscles, so that their own bodies betray them.

But the problem of thinking too much isn't limited to physical performers. Claude Steele, a professor of psychology at Stanford, studies the effects of performance anxiety on standardized-test scores. When Steele gave a large group of Stanford sophomores a set of questions from the Graduate Record Examination (GRE) and told the students that it would measure their innate intellectual ability, he found that the white students performed significantly better than their black counterparts. This discrepancy—commonly known as the achievement gap—conformed to a large body of data showing that minority students tend to score lower on a wide variety of standardized tests, from the SAT to the IQ test.

However, when Steele gave a separate group of students the same test but stressed that it was *not* a measure of intelligence —he told them it was merely a preparatory drill—the scores of the white and black students were virtually identical. The achievement gap had been closed. According to Steele, the disparity in test scores was caused by an effect that he calls stereotype threat. When black students are told that they are taking a

---

*A follow-up study found that instead of thinking about the mechanical details of the swing, experienced golfers should focus on general aspects of their intended movement, what psychologists call a holistic cue word. For instance, instead of contemplating something like the precise position of the wrist or elbow, the player should focus on a descriptive adjective, such as *smooth* or *balanced*. An experimental trial demonstrated that professional golfers who used these holistic cues did far better than golfers who consciously tried to control their strokes.

test to measure their intelligence, it brings to mind, rather force-fully, the ugly and untrue stereotype that blacks are less intelligent than whites. (Steele conducted his experiments soon after *The Bell Curve* was published, but the same effect also exists when women take a math test that supposedly measures "cognitive differences between the genders" or when white males are exposed to a stereotype about the academic superiority of Asians.) The Stanford sophomores were so worried about being viewed through the lens of a negative stereotype that they performed far below their abilities. "What you tend to see [during stereotype threat] is carefulness and second-guessing," Steele said. "When you go and interview them, you have the sense that when they are in the stereotype-threat condition they say to themselves, 'Look, I'm going to be careful here. I'm not going to mess things up.' Then, after having decided to take that strategy, they calm down and go through the test. But that's not the way to succeed on a standardized test. The more you do that, the more you will get away from the intuitions that help you, the quick processing. They think they did well, and they are trying to do well. But they are not."

The lesson of Renee Fleming, Jean Van de Velde, and these Stanford students is that rational thought can backfire. While reason is a powerful cognitive tool, it's dangerous to rely *exclusively* on the deliberations of the prefrontal cortex. When the rational brain hijacks the mind, people tend to make all sorts of decision-making mistakes. They hit bad golf shots and choose wrong answers on standardized tests. They ignore the wisdom of their emotions—the knowledge embedded in their dopamine neurons—and start reaching for things that they can explain. (One of the problems with feelings is that even when they are accurate, they can still be hard to articulate.) Instead of going with the option that feels the best, a person starts going with the option that *sounds* the best, even if it's a very bad idea.

# 1

When *Consumer Reports* tests a product, it follows a strict protocol. First, the magazine's staff assembles a field of experts. If they're testing family sedans, they rely on automotive experts; if audio speakers are being scrutinized, the staff members bring in people trained in acoustics. Then the magazine's staff gather all the relevant products in that category and try to hide the brand names. (This often requires lots of masking tape.) The magazine aspires to objectivity.

Back in the mid-1980s, *Consumer Reports* decided to conduct a taste test for strawberry jam. As usual, the editors invited several food experts, all of whom were "trained sensory panelists." These experts blindly sampled forty-five different jams, scoring each on sixteen different characteristics, such as sweetness, fruitiness, texture, and spreadability. The scores were then totaled, and the jams were ranked.

A few years later, Timothy Wilson, a psychologist at the University of Virginia, decided to replicate this taste test with his undergraduate students. Would the students have the same preferences as the experts? Did everybody agree on which strawberry jams tasted the best?

Wilson's experiment was simple: he took the first, eleventh, twenty-fourth, thirty-second, and forty-fourth best-tasting jams according to *Consumer Reports* and asked the students to rank them. In general, the preferences of the college students closely mirrored the preferences of the experts. Both groups thought Knott's Berry Farm and Alpha Beta were the two best-tasting brands, with Featherweight a close third. They also agreed that the worst strawberry jams were Acme and Sorrel Ridge. When Wilson compared the preferences of the students and the *Consumer Reports* panelists, he found that they had a statistical correlation of .55, which is rather impressive. When it comes to judging jam, we are all natural experts. Our brains are able to automatically pick out the products that provide us with the most pleasure.

But that was only the first part of Wilson's experiment. He repeated the jam taste test with a separate group of college students, only this time he asked them to explain *why* they preferred one brand over another. As they tasted the jams, the students filled out written questionnaires, which forced them to analyze their first impressions, to consciously explain their impulsive preferences. All this extra analysis seriously warped their jam judgment. The students now preferred Sorrel Ridge—the worst-tasting jam, according to *Consumer Reports*—to Knott's Berry Farm, which was the experts' favorite jam. The correlation plummeted to .11, which means that there was virtually no relationship between the rankings of the experts and the opinions of these introspective students.

Wilson argues that "thinking too much" about strawberry jam causes us to focus on all sorts of variables that don't actually matter. Instead of just listening to our instinctive preferences —the best jam is associated with the most positive feelings—our rational brains search for reasons to prefer one jam over another. For example, someone might notice that the Acme brand is particularly easy to spread, and so he'll give it a high ranking, even if he doesn't actually care about the spreadability of jam. Or a person might notice that Knott's Berry Farm jam has a chunky texture, which seems like a bad thing, even if she's never really thought about the texture of jam before. But having a chunky texture *sounds* like a plausible reason to dislike a jam, and so she revises her preferences to reflect this convoluted logic. People talk themselves into liking Acme jam more than the Knott's Berry Farm's product.

This experiment illuminates the danger of always relying on the rational brain. There is such a thing as too much analysis.

When you overthink at the wrong moment, you cut yourself off from the wisdom of your emotions, which are much better at assessing actual preferences. You lose the ability to know what you really want. And then you choose the worst strawberry jam.

WILSON WAS INTRIGUED by the strawberry-jam experiment. It seemed to contradict one of the basic tenets of Western thought, which is that careful self-analysis leads to wisdom. As Socrates famously said, "The unexamined life is not worth living." Socrates clearly didn't know about strawberry jam.

But perhaps food products are unique, since people are notoriously bad at explaining their own preferences. So Wilson came up with another experiment. This time he asked female college students to select their favorite poster. He gave them five options: a Monet landscape, a van Gogh painting of some purple lilies, and three humorous cat posters. Before making their choices, the subjects were divided into two groups. The first was the non-thinking group: they were instructed to simply rate each poster on a scale from 1 to 9. The second group had a tougher task: before they rated the posters, they were given questionnaires that asked them *why* they liked or disliked each of the five posters. At the end of the experiment, each of the subjects took her favorite poster home.

The two groups of women made very different choices. Ninety-five percent of the non-thinkers chose either the Monet or the van Gogh. They instinctively preferred the fine art. However, subjects who thought about their poster decisions first were almost equally split between the paintings and the humorous cat posters. What accounted for the difference? "When looking at a painting by Monet," Wilson writes, "most people generally have a positive reaction. When thinking about why they feel the way they do, however, what comes to mind and is easiest to verbalize might be that some of the colors are not very pleasing, and that the subject matter, a haystack, is rather boring." As a result, the women ended up selecting the funny feline posters, if only because those posters gave them more grist for their explanatory mill.

Wilson conducted follow-up interviews with the women a few weeks later to see which group had made the better decision. Sure enough, the members of the non-thinking group were much more satisfied with their choice of posters. While 75 percent of the people who had chosen cat posters regretted their selection, nobody regretted selecting the artistic poster. The women who listened to their emotions ended up making much better decisions than the women who relied on their reasoning powers. The more people thought about which posters they wanted, the more misleading their thoughts became. Self-analysis resulted in *less* self-awareness.

This isn't just a problem for insignificant decisions like choosing jam for a sandwich or selecting a cheap poster. People can also think too much about more important choices, like buying a home. As Ap Dijksterhuis, a psychologist at Radboud University, in the Netherlands, notes, when people are shopping for real estate, they often fall victim to a version of the strawberry-jam error, or what he calls a "weighting mistake." Consider two housing options: a three-bedroom apartment located in the middle of a city that would give you a ten-minute commute, and a five-bedroom McMansion in the suburbs that would result in a forty-five-minute commute. "People will think about this tradeoff for a long time," Dijksterhuis says, "and most of them will eventually choose the large house. After all, a third bathroom or extra bedroom is very important for when Grandma and Grandpa come over for Christmas, whereas driving two hours each day is really not that bad." What's interesting is the more time people spend deliberating, the more important that extra space becomes. They'll imagine all sorts of scenarios (a big birthday party, Thanksgiving dinner, another child) that turns the suburban house into a necessity. The lengthy commute, meanwhile, will seem less and less significant, at least when it's compared to the lure of an extra bathroom. But as Dijksterhuis points out, the reasoning is backward: "The additional bathroom is a completely superfluous asset for at least 362 or 363 days each year, whereas a long commute *does* become a burden after a while." For instance, a recent study found that when a person travels more than one hour in each direction, he or she has to make 40 percent more money in order to be as "satisfied with life" as a person with a short commute. Another study, led by Daniel Kahneman and the economist Alan Krueger, surveyed nine hundred workingwomen in Texas and found that commuting was, by far, the least pleasurable part of their day. And yet, despite these gloomy statistics, nearly 20 percent of American workers commute more than forty-five minutes each way. (More than 3.5 million Americans spend more than three hours each day traveling to and from work, and they're the fastest-growing category of commuter.) According to Dijksterhuis, all these people are making themselves miserable because they failed to properly weigh the relevant variables when they were choosing where to live. Just as strawberry-jam tasters who consciously analyzed their preferences were persuaded by irrelevant factors like spread ability and texture, the deliberative homeowners focused on less important details like square footage and number of bathrooms. (It's easier to consider quantifiable facts than future emotions, such as how you'll feel when you're stuck in a rush-hour traffic jam.) The prospective homeowners assumed a bigger house in the suburbs would make them happy, even if it meant spending an extra hour in the car every day. But they were wrong.

THE BEST WINDOW into this mental process—what's actually happening inside the brain when you talk yourself into choosing the wrong strawberry jam—comes from studies of the placebo effect. It's long been recognized that the placebo effect is extremely powerful; anywhere between 35 and 75 percent of people get better after receiving pretend medical treatments, such as sugar pills. A few years ago, Tor Wager, a neuroscientist at Columbia University, wanted to figure out why placebos were so effective. His experiment was brutally straightforward: he gave college students electric shocks while they were stuck in an fMRI machine. (The subjects were well compensated, at least by undergraduate standards.) Half of the people were then supplied with a fake pain-relieving cream. Even though the cream had no analgesic properties—it was just a hand moisturizer—people given the pretend cream said the shocks were significantly less painful. The placebo effect eased their suffering. Wager then imaged the specific parts of the brain that controlled this psychological process. He discovered that the placebo effect depended entirely on the prefrontal cortex, the center of reflective, deliberate thought. When people were told that they'd just received pain-relieving cream, their frontal lobes responded by inhibiting the activity of their emotional brain areas (like the insula) that normally respond to pain. Because people *expected* to experience less pain, they ended up experiencing less pain. Their predictions became self-fulfilling prophecies.

The placebo effect is a potent source of self-help. It demonstrates the power of the prefrontal cortex to modulate even the most basic bodily signals. Once this brain area comes up with reasons to experience less pain—the cream is supposed to provide pain relief—those reasons become powerful distortions. Unfortunately, the same rational brain areas responsible for temporarily reducing suffering also mislead us about many daily decisions. The prefrontal cortex can turn off pain signals, but it can also cause a person to ignore the feelings that lead to choosing the best poster. In these situations, conscious thoughts interfere with good decision-making.

Look, for example, at this witty little experiment. Baba Shiv, a neuroeconomist at Stanford, supplied a group of people with Sobe Adrenaline Rush, an "energy" drink that was supposed to make them feel more alert and energetic. (The drink contained a potent brew of sugar and caffeine that, the bottle promised, would impart "superior functionality.") Some participants paid full price for the drinks, while others were offered a discount. After drinking the product, participants were asked to solve a series of word puzzles. Shiv found that people who'd paid discounted prices consistently solved about 30 percent fewer puzzles than the people who'd paid full price for the drinks. The subjects were convinced that the stuff on sale was much less potent, even though all the drinks were identical. "We ran the study again and again, not sure if what we got had happened by chance or fluke," Shiv says. "But every time we ran it, we got the same results."

Why did the cheaper energy drink prove less effective? According to Shiv, consumers typically suffer from a version of the placebo effect. Since they *expect* cheaper goods to be less effective, they generally *are* less effective, even if the goods are identical to more expensive products. This is why brand-name aspirin works better than generic aspirin and why Coke tastes better than cheaper colas, even if most consumers can't tell the difference in blind taste tests. "We have these general beliefs about the world—for example, that cheaper products are of lower quality—and they translate into specific expectations about specific products," said Shiv. "Then, once these expectations are activated, they start to really impact our behavior." The rational brain distorts the sense of reality, so the ability to properly assess the alternatives is lost. Instead of listening to the trustworthy opinions generated by our emotional brains, we follow our own false assumptions.

Researchers at Caltech and Stanford recently lifted the veil on this strange process. Their experiment was organized like a wine-tasting. Twenty people sampled five cabernet sauvignons that

were distinguished solely by their retail prices, with bottles rang-
ing in cost from five dollars to ninety dollars. Although the people were told that all five wines were different, the scientists
weren't telling the truth: there were only three different wines.
This meant that the same wines often reappeared, but with different price labels. For example, the first wine offered during the
tasting—it was a bottle of a cheap California cabernet—was
labeled both as a five-dollar wine (its actual retail price) and as
a forty-five-dollar wine, a 900 percent markup. All of the red
wines were sipped by each subject inside an fMRI machine.

Not surprisingly, the subjects consistently reported that the
more expensive wines tasted better. They preferred the ninety-
dollar bottle to the ten-dollar bottle and thought the forty-five-
dollar cabernet was far superior to the five-dollar plonk. By conducting the winetasting inside an fMRI machine—the drinks
were sipped via a network of plastic tubes—the scientists could
see how the brains of the subjects responded to the different
wines. While a variety of brain regions were activated during the
experiment, only one brain region seemed to respond to the *price*
of the wine rather than the wine itself: the prefrontal cortex. In
general, more expensive wines made parts of the prefrontal cortex more excited. The scientists argue that the activity of this
brain region shifted the preferences of the winetasters, so that the
ninety-dollar cabernet seemed to taste better than the thirty-five-
dollar cabernet, even though they were actually the same wine.

Of course, the wine preferences of the subjects were clearly
nonsensical. Instead of acting like rational agents—getting the
most utility for the lowest possible price—they were choosing to
spend more money for an identical product. When the scientists
repeated the experiment with members of the Stanford University wine club, they got the same results. In a blind tasting, these
semi-experts were also misled by the made-up price tags. "We
don't realize how powerful our expectations are," says Antonio
Rangel, the neuroeconomist at Caltech who led the study. "They

can really modulate every aspect of our experience. And if our expectations are based on false assumptions"—like the assumption that more expensive wine tastes better—"they can be very misleading."

These experiments suggest that, in many circumstances, we could make better consumer decisions by knowing *less* about the products we are buying. When you walk into a store, you are besieged by information. Even purchases that seem simple can quickly turn into a cognitive quagmire. Look at the jam aisle. A glance at the shelves can inspire a whole range of questions. Should you buy the smooth-textured strawberry jam or the one with less sugar? Does the more expensive jam taste better? What about organic jam? (The typical supermarket contains more than two hundred varieties of jam and jelly.) Rational models of decision-making suggest that the way to find the best product is to take all of this information into account, to carefully analyze the different brands on display. In other words, a person should choose a jam with his or her prefrontal cortex. But this method can backfire. When we spend too much time thinking in the supermarket, we can trick ourselves into choosing the wrong things for the wrong reasons. That's why the best critics, from *Consumer Reports* to Robert Parker, always insist on blind comparisons. They want to avoid the deceptive thoughts that corrupt decisions. The prefrontal cortex isn't good at picking out jams or energy drinks or bottles of wine. Such decisions are like a golf swing: they are best done with the emotional brain, which generates its verdict automatically.

This "irrational" approach to shopping can save us lots of money. After Rangel and his colleagues finished their brain-imaging experiment, they asked the subjects to taste the five different wines again, only this time the scientists didn't provide any price information. Although the subjects had just listed the ninety-dollar wine as the most pleasant, they now completely reversed their preferences. When the tasting was truly blind, when the subjects were no longer biased by their prefrontal cortex, the cheapest wine got the highest ratings. It wasn't fancy, but it tasted the best.

2

If the mind were an infinitely powerful organ, a limitless super-computer without constraints, then rational analysis would always be the ideal decision-making strategy. Information would be an unqualified good. We would be foolish to ignore the omniscient opinions of the Platonic charioteer.

The biological reality of the brain, however, is that it's severely bounded, a machine subject to all sorts of shortcomings. This is particularly true of the charioteer, who is tethered to the prefrontal cortex. As the psychologist George Miller demonstrated in his famous essay "The Magical Number Seven, Plus or Minus Two," the conscious brain can only handle about seven pieces of data at any one moment. "There seems to be some limitation built into us by the design of our nervous systems, a limit that keeps our channel capacities in this general range," Miller wrote. While we can control these rational neural circuits—they think about what we tell them to think about—they constitute a relatively small part of the brain, just a few microchips within the vast mainframe of the mind. As a result, even choices that seem straightforward—like choosing a jam in the supermarket—can overwhelm the prefrontal cortex. It gets intimidated by all the jam data. And that's when bad decisions are made.

Consider this experiment. You're sitting in a bare room, with just a table and a chair. A scientist in a white lab coat walks in and says that he's conducting a study of long-term memory. The scientist gives you a seven-digit number to remember and asks you to walk down the hall to the room where your memory will be tested. On the way to the testing room, you pass a refreshment table for subjects taking part in the experiment. You are given a choice between a decadent slice of German chocolate cake and a bowl of fruit salad. What do you choose?

Now let's replay the experiment. You are sitting in the same room. The same scientist gives you the same explanation. The only difference is that instead of being asked to remember a seven-digit number, you are given only two numbers, a far easier mental task. You then walk down the hall and are given the same choice between cake and fruit.

You probably don't think the number of digits will affect your choice; if you choose the chocolate cake, it is because you want cake. But you'd be wrong. The scientist who explained the experiment was lying; this isn't a study of long-term memory, it's a study of self-control.

When the results from the two different memory groups were tallied, the scientists observed a striking shift in behavior. Fifty-nine percent of people trying to remember seven digits chose the cake, compared to only 37 percent of the two-digit subjects. Distracting the brain with a challenging memory task made a person much more likely to give in to temptation and choose the calorie-dense dessert. (The premise is that German chocolate cake is to adults what marshmallows are to four-year-olds.) The subjects' self-control was overwhelmed by five extra numbers.

Why did the two groups behave so differently? According to the Stanford scientists who designed the experiment, the effort required to memorize seven digits drew cognitive resources away from the part of the brain that normally controls emotional urges. Because working memory and rationality share a common cortical source—the prefrontal cortex—a mind trying to remember lots of information is less able to exert control over its impulses. The substrate of reason is so limited that a few extra digits can become an extreme handicap.

The shortcomings of the prefrontal cortex aren't apparent only when memory-storage capacity is exceeded. Other studies have shown that a slight drop in blood-sugar levels can also inhibit self-control, since the frontal lobes require lots of energy in order to function. Look, for example, at this experiment led by Roy Baumeister, a psychologist at Florida State University. The experiment began with a large group of undergraduates performing a mentally taxing activity that involved watching a video while ignoring the text of random words scrolling on the bottom of the screen. (It takes a conscious effort to *not* pay attention to salient stimuli.) The students were then offered some lemonade. Half of them got lemonade made with real sugar, and the other half got lemonade made with a sugar substitute. After giving the glucose time to enter the bloodstream and perfuse the brain (about fifteen minutes), Baumeister had the students make decisions about apartments. It turned out that the students who were given the drink without real sugar were significantly more likely to rely on instinct and intuition when choosing a place to live, even if that led them to choose the wrong places. The reason, according to Baumeister, is that the rational brains of these students were simply to exhausted to think. They'd needed a restorative sugar fix, and all they'd gotten was Splenda. This research can also help explain why we get cranky when we're hungry and tired: the brain is less able to suppress the negative emotions sparked by small annoyances. A bad mood is really just a run-down prefrontal cortex.

The point of these studies is that the flaws and foibles of the rational brain—the fact that it's an imperfect piece of machinery—are constantly affecting our behavior, leading us to make decisions that seem, in retrospect, quite silly. These mistakes extend far beyond poor self-control. In 2006, psychologists at the University of Pennsylvania decided to conduct an experiment with M&M's in an upscale apartment building. One day, they left out a bowl of the chocolate candies and a small scoop. The next day they refilled the bowl with M&M's but placed a much larger scoop beside it. The result would not surprise anyone who has ever finished a Big Gulp soda or a supersize serving of Mc-Donald's fries: when the scoop size was increased, people took 66 percent more M&M's. Of course, they could have taken just as many candies on the first day; they simply would have had to use a few more scoops. But just as larger serving sizes cause us to eat more, the larger scoop made the residents more gluttonous.

The real lesson of the candy scoop, however, is that people are terrible at measuring stuff. Instead of counting the number of M&M's they eat, they count the number of scoops. The scientists found that most people took a single scoop and ended up consuming however many candies that scoop happened to contain. The same thing happens when people sit down to dinner: they tend to eat whatever is on their plates. If the plate is twice as large (and American serving sizes have grown 40 percent in the last twenty-five years), they'll still polish it off. As an example, a study done by Brian Wansink, a professor of marketing at Cornell, used a bottomless bowl of soup—there was a secret tube that kept on refilling the bowl with soup from below—to demonstrate that how much people eat is largely dependent on serving size. The group with the bottomless bowls ended up consuming nearly 70 percent more soup than the group with normal bowls.

Economists call this sleight of mind mental accounting, since people tend to think about the world in terms of specific accounts, such as scoops of candy or bowls of soup or lines on a budget. While these accounts help people think a little faster —it's easier to count scoops than actual M&M's—they also distort decisions. Richard Thaler, an economist at the University of Chicago, was the first to fully explore the consequences of this irrational behavior. He came up with a simple set of questions that demonstrate mental accounting at work:

Imagine that you have decided to see a movie and have paid the admission price of $10 per ticket. As you enter the theater, you discover that you have lost the ticket. The seat was not marked, and the ticket cannot be recovered. Would you pay $10 for another ticket?

When Thaler conducted this survey, he found that only 46 percent of people would buy another movie ticket. However, when he asked a closely related question, he got a completely different response.

Imagine that you have decided to see a movie where admission is $10, but you have not yet bought the ticket. As you walk to the theater, you discover that you have lost a $10 bill. Would you still pay $10 for a ticket to the movie?

Although the value of the loss in both scenarios is the same —people were still losing ten dollars—88 percent of people said they would now buy a movie ticket. Why the drastic shift? According to Thaler, going to a movie is normally viewed as a transaction in which the cost of a ticket is exchanged for the experience of seeing a movie. Buying a second ticket makes the movie seem too expensive, since a single ticket now "costs" twenty dollars. In contrast, the loss of the cash is not posted to the mental account of the movie, so no one minds forking over another ten bucks.

Of course, this is woefully inconsistent behavior. After losing tickets, most of us become tightwads; when we lose merely cash, we remain spendthrifts. These contradictory decisions violate an important principle of classical economics, which assumes that a dollar is always a dollar. (Money is supposed to be perfectly fungible.) But because the brain engages in mental accounting, we end up treating our dollars very differently. For example, when Thaler asked people whether they would drive twenty minutes out of their way to save five dollars on a fifteen-dollar calculator, 68 percent of respondents said yes. However, when he asked people whether they would drive twenty minutes out of their way to save five dollars on a $125 leather jacket, only 29 percent said they would. Their driving decisions depended less on the absolute amount of money involved (five dollars) than on the particular mental account in which the decision was placed. If the savings activated a mental account with a minuscule amount of money—like buying a cheap calculator—then they were compelled to drive across town. But that same five dollars seems irrelevant when part of a much larger purchase. This principle also explains why car dealers are able to tack on unwanted and expensive extras and why luxury hotels can get away with charging six dollars for a can of peanuts. Because these charges are only small parts of much bigger purchases, we end up paying for things that we wouldn't normally buy.

The brain relies on mental accounting because it has such limited processing abilities. As Thaler notes, "These thinking problems come from the fact that we have a slow, erratic CPU [central processing unit] and the fact that we're busy." Since the prefrontal cortex can handle only about seven things at the same time, it's constantly trying to "chunk" stuff together, to make the complexity of life a little more manageable. Instead of thinking about each M&M, we think about the scoops. Instead of counting every dollar we spend, we parcel our dollars into particular purchases, like cars. We rely on misleading shortcuts because we lack the computational power to think any other way.

# 3

The history of Western thought is so full of paeans to the virtues of rationality that people have neglected to fully consider its limitations. The prefrontal cortex, it turns out, is easy to hoodwink. All it takes is a few additional digits or a slightly bigger candy scoop, and this rational brain region will start making irrational decisions.

A few years ago, a group of MIT economists led by Dan Ariely decided to conduct an auction with their business-school graduate students. (The experiment was later conducted on executives and managers at the MIT Executive Education Program, with similar results.) The researchers were selling a motley group of items, from a fancy bottle of French wine to a cordless keyboard to a box of chocolate truffles. The auction, however, came with a twist: before the students could bid, they were asked to write down the last two digits of their Social Security numbers. Then they were supposed to say whether or not they would be willing to pay that numerical amount for each of the products. For instance, if the last two digits of the number were 55, then the student would have to decide whether the bottle of wine or the cordless keyboard was worth $55. Finally, the students were instructed to write down the maximum amount they were willing to pay for the various items.

If people were perfectly rational agents, if the brain weren't so bounded, then writing down the last two digits of their Social Security numbers should have no effect on their auction bids. In other words, a student whose Social Security number ended with a low-value figure (such as 10) should be willing to pay roughly the same price as someone with a high-value figure (such as 90). But that's not what happened. For instance, look at the bidding for the cordless keyboard. Students with the highest-ending Social Security numbers (80–99) made an average bid of fifty-six dollars. In contrast, students with the lowest-ending numbers (1–20) made an average bid of a paltry sixteen dollars. A similar trend held for every single item. On average, students with higher numbers were willing to spend 300 *percent* more than those with low numbers. All of the business students realized, of course, that the last two digits of their Social Security numbers were completely irrelevant. Such a thing shouldn't influence their bids. And yet, it clearly did.

This is known as the anchoring effect, since a meaningless anchor—in this case, a random number—can have a strong impact on subsequent decisions.* While it's easy to mock the irrational bids of the business students, the anchoring effect is actually a common consumer mistake. Consider the price tags in a car dealership. Nobody actually pays the prices listed in bold black ink on the windows. The inflated sticker is merely an anchor that allows the car salesperson to make the real price of the car seem like a better deal. When a person is offered the inevitable discount, the prefrontal cortex is convinced that the car is a bargain.

In essence, the anchoring effect is about the brain's spectacular inability to dismiss irrelevant information. Car shoppers should ignore the manufacturers' suggested retail prices, just as MIT grad students should ignore their Social Security numbers. The problem is that the rational brain isn't good at disregarding facts, even when it knows those facts are useless. And so, if someone is looking at a car, the sticker price serves as a point of comparison, even though it's merely a gimmick. And when a person in the MIT experiment is making a bid on a cordless keyboard, she can't help but tender an offer that takes her Social Security number into account, simply because that number has already been placed into the pertinent decision-making ledger. The random digits are stuck in her prefrontal cortex, occupying valuable cognitive space. As a result, they become a starting point when

---

*Daniel Kahneman first demonstrated the anchoring effect in an experiment known as the United Nations game. He asked people to estimate the percentage of African countries in the United Nations. Before they guessed, a random number was generated—directly in front of the participants—by spinning a roulette wheel. As you might imagine, people who saw higher numbers on the roulette wheel generated significantly higher guesses for the percentage of African countries in the United Nations than those who saw lower numbers.

she thinks about how much she's willing to pay for a computer accessory. "You know you're not supposed to think about these meaningless numbers," Ariely says. "But you just can't help it."

The fragility of the prefrontal cortex means that we all have to be extremely vigilant about not paying attention to unnecessary information. The anchoring effect demonstrates how a single additional fact can systematically distort the reasoning process. Instead of focusing on the important variable—how much is that cordless keyboard really worth?—we get distracted by some meaningless numbers. And then we spend too much money.

This cortical flaw has been exacerbated by modernity. We live in a culture that's awash in information; it's the age of Google, cable news, and free online encyclopedias. We get anxious whenever we are cut off from all this knowledge, as if it's impossible for anyone to make a decision without a search engine. But this abundance comes with some hidden costs. The main problem is that the human brain wasn't designed to deal with such a surfeit of data. As a result, we are constantly exceeding the capacity of our prefrontal cortices, feeding them more facts and figures than they can handle. It's like trying to run a new computer program on an old machine; the antique microchips try to keep up, but eventually they fizzle out.

In the late 1980s, the psychologist Paul Andreassen conducted a simple experiment on MIT business students. (Those poor students at MIT's Sloan School of Management are very popular research subjects. As one scientist joked, "They're like the fruit fly of behavioral economics.") First, Andreassen let each of the students select a portfolio of stock investments. Then he divided the students into two groups. The first group could see only the changes in the prices of their stocks. They had no idea why the share prices rose or fell and had to make their trading decisions based on an extremely limited amount of data. In contrast, the second group was given access to a steady stream of financial information. They could watch CNBC, read the *Wall Street Journal*, and consult experts for the latest analysis of market trends.

So which group did better? To Andreassen's surprise, the group with less information ended up earning more than twice as much as the well-informed group. Being exposed to extra news was distracting, and the high-information students quickly became focused on the latest rumors and insider gossip. (Herbert Simon said it best: "A wealth of information creates a poverty of attention.") As a result of all the extra input, these students engaged in far more buying and selling than the low-information group. They were convinced that all their knowledge allowed them to anticipate the market. But they were wrong.

The dangers of too much information aren't confined to investors. In another study, college counselors were given a vast amount of information about a group of high school students. The counselors were then asked to predict the grades of these kids during their freshman year in college. The counselors had access to high school transcripts, test scores, the results of personality and vocational tests, and application essays from the students. They were even granted personal interviews so that they could judge the "academic talents" of the students in person. With access to all of this information, the counselors were extremely confident that their judgments were accurate.

The counselors were competing against a rudimentary mathematical formula composed of only two variables: the high school grade point average of the student and his or her score on a single standardized test. Everything else was deliberately ignored. Needless to say, the predictions made by the formula were far more accurate than the predictions made by the counselors. The human experts had looked at so many facts that they lost track of which facts were actually important. They subscribed to illusory correlations ("She wrote a good college essay, so she'll write good essays in college") and were swayed by irrelevant details ("He had such a nice smile"). While the extra information considered by the counselors made them extremely confident, it actually led to worse predictions. Knowledge has diminishing returns, right up until it has negative returns.

This is a counterintuitive idea. When making decisions, people almost always assume that more information is better. Modern corporations are especially beholden to this idea and spend a fortune trying to create "analytic workspaces" that "maximize the informational potential of their decision-makers." These managerial clichés, plucked from the sales brochures of companies such as Oracle and Unisys, are predicated on the assumptions that executives perform better when they have access to more facts and figures and that bad decisions are a result of ignorance.

But it's important to know the limitations of this approach, which are rooted in the limitations of the brain. The prefrontal cortex can handle only so much information at any one time, so when a person gives it too many facts and then asks it to make a decision based on the facts that *seem* important, that person is asking for trouble. He is going to buy the wrong items at Wal-Mart and pick the wrong stocks. We all need to know about the innate frailties of the prefrontal cortex so that we don't undermine our decisions.

BACK PAIN IS a medical epidemic. The numbers are sobering: there's a 70 percent chance that at some point in your life, you'll suffer from it. There's a 30 percent chance that you've suffered from severe back pain in the last thirty days. At any given time, about 1 percent of working-age Americans are completely incapacitated by their lower lumbar regions. Treatment is expensive (more than $26 billion a year) and currently accounts for about 3 percent of total health-care spending. If workers' compensation and disability payments are taken into account, the costs are far higher.

When doctors first started to encounter a surge in patients with back pain—the beginning of the epidemic is generally dated to the late 1960s—they had few answers. The lower back is an exquisitely complicated body area, full of tiny bones, ligaments, spinal discs, and minor muscles. And then there's the spinal cord itself, a thick sheath of sensitive nerves that can be easily upset. There are so many moving parts in the back that doctors had difficulty figuring out what exactly was responsible for the pain. Without a definitive explanation, doctors typically sent patients home with a prescription for bed rest.

But this simple treatment plan was extremely effective. Even when nothing was done to the lower back, about 90 percent of patients with back pain managed to get better within seven weeks. The body healed itself, the inflammation subsided, the nerves relaxed. These patients went back to work and pledged to avoid the sort of physical triggers that had caused the pain in the first place.

Over the next few decades, this hands-off approach to back pain remained the standard medical treatment. Although the vast majority of patients didn't receive a specific diagnosis of what caused the pain—the suffering was typically parceled into a vague category such as "lower lumbar strain"—they still managed to experience significant improvements within a short period of time. "It was a classic case of medicine doing best by doing least," says Dr. Eugene Carragee, a professor of orthopedic surgery at Stanford. "People got better without real medical interventions because doctors didn't know how to intervene."

That all changed with the introduction of magnetic resonance imaging (MRI) in the late 1980s. Within a few years, the MRI machine became a crucial medical tool. It allowed doctors to look, for the first time, at stunningly accurate images of the interior of the body. MRI machines use powerful magnets to make protons in the flesh shift ever so slightly. Different tissues react in slightly different ways to this atomic manipulation; a computer then translates the resulting contrasts into high-resolution images. Thanks to the precise pictures produced by the machine, doctors no longer needed to imagine the layers of matter underneath the skin. They could see everything.

The medical profession hoped that the MRI would revolutionize the treatment of lower back pain. Since doctors could finally image the spine and surrounding soft tissue in lucid detail, they figured they'd be able to offer precise diagnoses of what was causing the pain, locating the aggravated nerves and structural problems. This, in turn, would lead to better medical care.

Unfortunately, MRIs haven't solved the problem of back pain. In fact, the new technology has probably made the problem worse. The machine simply sees too much. Doctors are overwhelmed with information and struggle to distinguish the significant from the irrelevant. Take, for example, spinal disc abnormalities. While x-rays can reveal only tumors and problems with the vertebral bones, MRIs can image spinal discs—the supple buffers between the vertebrae—in meticulous detail. After the imaging machines were first introduced, the diagnoses of various disc abnormalities began to skyrocket. The MRI pictures certainly looked bleak: people with pain seemed to have seriously degenerated discs, which everyone assumed caused inflammation of the local nerves. Doctors began administering epidurals to quiet the pain, and if the pain persisted, they would surgically remove the apparently offending disc tissue.

The vivid images, however, were misleading. Those disc abnormalities are seldom the cause of chronic back pain. In a 1994 study published in the *New England Journal of Medicine,* a group of researchers imaged the spinal regions of ninety-eight people who had no back pain or back-related problems. The pictures were then sent to doctors who didn't know that the patients weren't in pain. The result was shocking: the doctors reported that two-thirds of these normal patients exhibited "serious problems" such as bulging, protruding, or herniated discs. In 38 percent of these patients, the MRI revealed multiple damaged discs. Nearly 90 percent of these patients exhibited some form of "disc degeneration." These structural abnormalities are often used to justify surgery, and yet nobody would advocate surgery for people without pain. The study concluded that, in most cases, "The discovery by MRI of bulges or protrusions in people with low back pain may be coincidental."

In other words, seeing everything made it harder for the doctors to know what they should be looking at. The very advantage of MRI—its ability to detect tiny defects in tissue—turned out to be a liability, since many of the so-called defects were actually normal parts of the aging process. "A lot of what I do is educate people about what their MRIs are showing," says Dr. Sean Mackey, a professor at the Stanford School of Medicine and associate director of the hospital's pain-management division. "Doctors and patients get so fixated on these slight disc problems, and then they stop thinking about other possible causes for the pain. I always remind my patients that the only perfectly healthy spine is the spine of an eighteen-year-old. Forget about your MRI. What it's showing you is probably not important."

The mistaken explanations for back pain triggered by MRIs inevitably led to an outbreak of bad decisions. A large study published in the *Journal of the American Medical Association* (*JAMA*) randomly assigned 380 patients with back pain to undergo two different types of diagnostic analysis. One group received x-rays. The other group got diagnosed using MRIs, which gave the doctors much more information about the underlying anatomy.

Which group fared better? Did better pictures lead to better treatments? There was no difference in patient outcome: the vast majority of people in both groups got better. More information didn't lead to less pain. But stark differences emerged when the study looked at *how* the different groups were treated. Nearly

50 percent of MRI patients were diagnosed with some sort of disc abnormality, and this diagnosis led to intensive medical interventions. The MRI group had more doctor visits, more injections, more physical therapy, and were more than twice as likely to undergo surgery. These additional treatments were very expensive, and they had no measurable benefit.

This is the danger of too much information: it can actually interfere with understanding. When the prefrontal cortex is overwhelmed, a person can no longer make sense of the situation. Correlation is confused with causation, and people make theories out of coincidences. They latch on to medical explanations, even when the explanations don't make very much sense. MRIs make it easy for doctors to see all sorts of disc "problems," and so they reasonably conclude that these structural abnormalities are causing the pain. They're usually wrong.

Medical experts are now encouraging doctors *not* to order MRIs when evaluating back pain. A recent report in the *New England Journal of Medicine* concluded that MRIs should be used to image the back only under specific clinical circumstances, such as when doctors are examining "patients for whom there is a strong clinical suggestion of underlying infection, cancer, or persistent neurologic deficit." In the latest clinical guidelines issued by the American College of Physicians and the American Pain Society, doctors were "strongly recommended . . . not to obtain imaging or other diagnostic tests in patients with nonspecific low back pain." In too many cases, the expensive tests proved worse than useless. All of the extra detail just got in the way. The doctors performed better with less information.

And yet, despite these clear medical recommendations, MRIs continue to be routinely prescribed by physicians trying to diagnose causes of back pain. The addiction to information can be hard to break. A 2003 report in *JAMA* found that even when doctors were aware of medical studies criticizing the use of MRI, they still believed that imaging was necessary for their own patients. They wanted to find a *reason* for the pain so that the suffering could be given a clear anatomical cause, which could then be fixed with surgery. It didn't seem to matter that these reasons weren't empirically valid, or that the disc problems seen by MRI machines don't actually cause most cases of lower back pain. More data was seen as an unqualified good. The doctors thought it would be irresponsible not to conduct all of the relevant diagnostic tests. After all, wasn't that the rational thing to do? And shouldn't doctors always try to make rational decisions?

The problem of diagnosing the origins of back pain is really just another version of the strawberry-jam problem. In both cases, the rational methods of decision-making cause mistakes. Sometimes, more information and analysis can actually constrict thinking, making people understand less about what's really going on. Instead of focusing on the most pertinent variable—the percentage of patients who get better and experience less pain—doctors got sidetracked by the irrelevant MRI pictures.

When it comes to treating back pain, this wrong-headed approach comes with serious costs. "What's going on now is a disgrace," says Dr. John Sarno, a professor of clinical rehabilitation at New York University Medical Center. "You have well-meaning doctors making structural diagnoses despite a serious lack of evidence that these abnormalities are really causing the chronic pain. But they have these MRI pictures and the pictures seem so convincing. It's amazing how perfectly intelligent people will make foolish decisions if you give them lots of irrelevant stuff to consider."

The powers of the Platonic charioteer are fragile. The prefrontal cortex is a magnificent evolutionary development, but it must be used carefully. It can monitor thoughts and help evaluate emotions, but it can also paralyze, making a person forget the words to an aria or lose a trusty golf swing. When someone falls into the trap of spending too much time thinking about fine-art posters or about the details of an MRI image, the rational brain is being used in the wrong way. The prefrontal cortex can't handle so much complexity by itself.

So far, this book has been about the brain's dueling systems. We've seen how both reason and feeling have important strengths and weaknesses, and how, as a result, different situations require different cognitive strategies. How we decide should depend on what we are deciding.

But before we learn how to take full advantage of our varied mental tools, we are going to explore a separate realm of decision-making. As it happens, some of our most important decisions are about how we treat other people. The human being is a social animal, endowed with a brain that shapes social behavior. By understanding how the brain makes these decisions, we can gain insight into one of most unique aspects of human nature: morality.

# 6

# The Moral Mind

When John Wayne Gacy was a child, he liked to torture animals. He caught mice in a wire trap and then cut them open with scissors while they were still alive. The blood and guts didn't bother him. Neither did the squeals. Sadism was entertaining.

This streak of cruelty was one of the few noteworthy facts of Gacy's childhood. In just about every other respect, his early years were utterly normal. He grew up in the middle-class suburbs of Chicago, where he was a Boy Scout and delivered the local newspaper. He got good grades in school but didn't want to go to college. When his high school classmates were later asked what they remembered about Gacy, most couldn't remember anything. He blended in with the crowd.

Gacy grew up to become a successful construction contractor and a pillar of the community. He liked to throw big summer barbecues, grill hot dogs and hamburgers and invite the neighbors over. He dressed up as a clown for kids in the hospital and was active in local politics. The local chamber of commerce voted him Man of the Year. He was a typical suburban husband.

The normalcy, however, was a carefully crafted lie. One day, Gacy's wife noticed a pungent odor coming from the crawlspace underneath their house. It was probably just a dead rodent, Gacy said, or maybe a sewage leak. He bought a fifty-pound bag of lime and tried to erase the smell. But the smell wouldn't go away. Gacy filled in the space with concrete. The smell still wouldn't go away. There was something bad underneath those floorboards.

The smell was rotting bodies. On March 12, 1980, John Wayne Gacy was convicted of murdering thirty-three boys. He paid the boys for sex, and if something went awry with the transaction, he would kill them in his living room. Sometimes he killed one after he raised his price. Or if he thought the boy might tell somebody. Or if he didn't have enough cash in his wallet. Sometimes he killed a boy because it seemed like the easiest thing to do. He'd put a sock in his mouth, strangle him with a rope, and get rid of the corpse in the middle of the night. When the cops finally searched Gacy's home, they found skeletons everywhere: underneath his garage, in the basement, in the backyard. The graves were shallow, just a few inches deep.

# 1

John Wayne Gacy was a psychopath. Psychiatrists estimate that about 25 percent of the prison population have psychopathic tendencies, but the vast majority of these people will never kill anybody. While psychopaths are prone to violence—especially when the violence is used to achieve a goal, like satisfying a sexual desire—their neurological condition is best defined in terms of a specific brain malfunction: psychopaths make poor—sometimes disastrous—moral choices.

At first glance, it seems strange to think of psychopaths as decision-makers. We tend to label people like John Wayne Gacy as monsters, horrifying examples of humanity at its most inhuman. But every time Gacy murdered a boy, killing without the slightest sense of unease, he was making a decision. He was willingly violating one of the most ancient of moral laws: *thou shalt not kill.* And yet Gacy felt no remorse; his conscience was clean, and he slept like a baby.

Psychopaths shed light on a crucial subset of decision-making that's referred to as morality. Morality can be a squishy, vague concept, and yet, at its simplest level, it's nothing but a series of choices about how we treat other people. When you act in a moral manner—when you recoil from violence, treat others fairly, and help strangers in need—you are making decisions that take people besides yourself into account. You are thinking about the feelings of others, sympathizing with their states of mind. This is what psychopaths can't do.

What causes this profound deficit? On most psychological tests, psychopaths appear perfectly normal. The working memory isn't impaired, they use language normally, and they don't have reduced attention spans. In fact, several studies have found that psychopaths have above-average IQs and reasoning abilities. Their logic is impeccable. But this intact intelligence conceals a devastating disorder: psychopaths are dangerous because they have damaged emotional brains.

Just look at Gacy. According to a court-appointed psychiatrist, Gacy seemed incapable of experiencing regret, sadness, or joy. He never lost his temper or got particularly angry. Instead, his inner life consisted entirely of sexual impulses and ruthless rationality. He felt nothing, but planned everything. (Gacy's meticulous criminal preparations are what allowed him to evade the police for so long.) Alec Wilkinson, a journalist who spent hours interviewing Gacy on death row, described his eerily detached demeanor in *The New Yorker:*

> [Gacy] appears to have no inner being. I often had the feeling that he was like an actor who had created a role and polished it

so carefully that he had become the role and the role had become him. In support of his innocence, he often says things that are deranged in their logic, but he says them so calmly that he appears to be rational and reasonable . . . Compared to other murderers at the prison, Gacy seemed tranquil.

This sort of emotional emptiness is typical of psychopaths. When normal people are shown staged videos of strangers being subjected to pain—for example, receiving powerful electric shocks—they automatically generate visceral emotional reactions. Their hands start to sweat and their blood pressure surges. But psychopaths feel nothing. It's as if they were watching a blank screen. Most people react differently to emotionally charged verbs such as *kill* or *rape* than they do to neutral words such as *sit* or *walk,* but that's not the case with psychopaths. For them, the words all seem equivalent. When normal people tell lies, they exhibit the classic symptoms of nervousness; lie detectors work by measuring these signals. But psychopaths are able to consistently fool the machines. Dishonesty doesn't make them anxious because nothing makes them anxious. They can lie with impunity. When criminologists looked at the most violent wife batterers, they discovered that as those men became more and more aggressive, their blood pressure and pulse rates actually *dropped.* The acts of violence had a calming effect.

"Psychopaths have a fundamental emotional disorder," says James Blair, a cognitive psychologist at the National Institute of Mental Health and coauthor of *The Psychopath: Emotion and the Brain.* "You know when you see a scared face in a movie and that makes you automatically feel scared, too? Well, psychopaths don't feel that. It's like they don't understand what's going on. This lack of emotion is what causes their dangerous behavior. They are missing the primal emotional cues that the rest of us use as guides when making moral decisions."

When you peer inside the psychopathic brain, you can see this absence of emotion. After being exposed to fearful facial expressions, the emotional parts of the normal human brain show increased levels of activation. So do the cortical areas responsible for recognizing faces. As a result, a frightened face becomes a frightening sight; we naturally internalize the feelings of others. The brain of a psychopath, however, responds to these fearful faces with an utter lack of interest. The emotional areas are unperturbed, and the facial-recognition system is even *less* interested in fearful faces than it is in perfectly blank stares. The psychopath's brain is bored by expressions of terror.

While the anatomy of evil remains incomplete, neuroscientists are beginning to identify the specific deficits that define the psychopathic brain. The main problem seems to be a broken amygdala, a brain area responsible for propagating aversive emotions such as fear and anxiety. As a result, psychopaths never feel bad when they make other people feel bad. Aggression doesn't make them nervous. Terror isn't terrifying. (Brain-imaging studies have demonstrated that the human amygdala is activated when a person merely *thinks* about committing a "moral transgression.") This emotional void means that psychopaths never learn from their adverse experiences; they are four times more likely than other prisoners to commit crimes after being released. For a psychopath on parole, there is nothing inherently wrong with violence. Hurting someone else is just another way of getting what he wants, a perfectly reasonable way to satisfy desires. The absence of emotion makes the most basic moral concepts incomprehensible. G. K. Chesterton was right: "The madman is not the man who has lost his reason. The madman is the man who has lost everything except his reason."

AT FIRST GLANCE, the connection between morality and the emotions might be a little unnerving. Moral decisions are supposed to rest on a firm logical and legal foundation. Doing the right thing means carefully weighing competing claims, like a *dis*passionate judge. These aspirations have a long history. The luminaries of the Enlightenment, such as Leibniz and Descartes, tried to construct a moral system entirely free of feelings. Immanuel Kant argued that doing the right thing was merely a consequence of acting rationally. Immorality, he said, was a result of illogic. "The oftener and more steadily we reflect" on our moral decisions, Kant wrote, the more moral those decisions become. The modern legal system still subscribes to this antiquated set of assumptions and pardons anybody who demonstrates a "defect in rationality"—these people are declared legally insane—since the rational brain is supposedly responsible for distinguishing between right and wrong. If you can't reason, then you shouldn't be punished.

But all of these old conceptions of morality are based on a fundamental mistake. Neuroscience can now see the substrate of moral decisions, and there's nothing rational about it. "Moral judgment is like aesthetic judgment," writes Jonathan Haidt, a psychologist at the University of Virginia. "When you see a painting, you usually know instantly and automatically whether you like it. If someone asks you to explain your judgment, you confabulate . . . Moral arguments are much the same: Two people feel strongly about an issue, their feelings come first, and their reasons are invented on the fly, to throw at each other."

Kant and his followers thought the rational brain acted like a scientist: we used reason to arrive at an accurate view of the world. This meant that morality was based on objective values; moral judgments described moral facts. But the mind doesn't work this way. When you are confronted with an ethical dilemma, the unconscious automatically generates an emotional reaction. (This is what psychopaths can't do.) Within a few milliseconds, the brain has made up its mind; you know what is right and what is wrong. These moral instincts aren't rational

—they've never heard of Kant—but they are an essential part of what keep us all from committing unspeakable crimes.

It's only at this point—*after* the emotions have already made the moral decision—that those rational circuits in the prefrontal cortex are activated. People come up with persuasive reasons to justify their moral intuition. When it comes to making ethical decisions, human rationality isn't a scientist, it's a *lawyer*. This inner attorney gathers bits of evidence, post hoc justifications, and pithy rhetoric in order to make the automatic reaction seem reasonable. But this reasonableness is just a façade, an elaborate self-delusion. Benjamin Franklin said it best in his autobiography: "So convenient a thing it is to be a reasonable creature, since it enables one to find or make a reason for everything one has a mind to do."

In other words, our standard view of morality—the philosophical consensus for thousands of years—has been exactly backward. We've assumed that our moral decisions are the by-products of rational thought, that humanity's moral rules are founded in such things as the Ten Commandments and Kant's categorical imperative. Philosophers and theologians have spilled lots of ink arguing about the precise logic of certain ethical dilemmas. But these arguments miss the central reality of moral decisions, which is that logic and legality have little to do with anything.

Consider this moral scenario, which was first invented by Haidt. Julie and Mark are siblings vacationing together in the south of France. One night, after a lovely day spent exploring the local countryside, they share a delicious dinner and a few bottles of red wine. One thing leads to another and Julie and Mark decide to have sex. Although she's on the pill, Mark uses a condom just in case. They enjoy themselves very much, but decide not to have sex again. The siblings promise to keep the one-night affair secret and discover, over time, that having sex has brought them even closer together. Did Julie and Mark do something wrong?*

If you're like most people, your first reaction is that the brother and sister committed a grave sin. What they did was very wrong. When Haidt asks people to explain their harsh moral judgments, the most common reasons given are the risk of having kids with genetic abnormalities and the possibility that sex will damage the sibling relationship. At this point, Haidt politely points out that Mark and Julie used two types of birth control and that having sex actually improved their relationship. But the facts of the case don't matter. Even when their arguments are disproved, people still cling to the belief that having sex with one's brother or sister is somehow immoral.

"What happens in the experiment," Haidt says, "is [that] people give a reason [why the sex is wrong]. When that reason is stripped from them, they give another reason. When the new reason is stripped from them, they reach for *another* reason." Eventually, of course, people run out of reasons: they've exhausted their list of moral justifications. The rational defense is forced to rest its case. That's when people start saying things like "Because it's just wrong to have sex with your sister" or "Because it's disgusting, that's why!" Haidt calls this state "moral dumbfounding." People know something seems morally wrong —sibling sex is a terrible idea—but no one can rationally defend the verdict. According to Haidt, this simple story about sibling sex illuminates the two separate processes that are at work when we make moral decisions. The emotional brain generates the verdict. It determines what is wrong and what is right. In the case of Julie and Mark, it refuses to believe that having sex with a sibling is morally permissible, no matter how many forms of birth control are used. The rational brain, on the other hand, *explains*

---

*Haidt's other scenarios involve a woman who uses the American flag to clean her bathroom and a family that eats their dog after it has been killed by a car.

the verdict. It provides reasons, but those reasons all come after the fact.

This is why psychopaths are so dangerous: they are missing the emotions that guide moral decisions in the first place. There's a dangerous void where their feelings are supposed to be. For people like Gacy, sin is always intellectual, never visceral. As a result, a psychopath is left with nothing but a rational lawyer inside his head, willing to justify any action. Psychopaths commit violent crimes because their emotions never tell them not to.

## 2

Moral decisions are a unique kind of decision. When you're picking out products in the grocery store, searching for the best possible strawberry jam, you are trying to maximize your own enjoyment. You are the only person that matters; it is your sense of pleasure that you are trying to please. In this case, selfishness is the ideal strategy. You should listen to those twitchy cells in the orbitofrontal cortex that tell you what you really want.

However, when you are making a moral decision, this egocentric strategy backfires. Moral decisions require taking *other people* into account. You can't act like a greedy brute or let your anger get out of control; that's a recipe for depravity and jail time. Doing the right thing means thinking about everybody else, using the emotional brain to mirror the emotions of strangers. Selfishness needs to be balanced by some selflessness.

The evolution of morality required a whole new set of decision-making machinery. The mind needed to evolve some structures that would keep it from hurting other people. Instead of just seeking more pleasure, the brain had to become sensitive to the pain and plight of strangers. The new neural structures that developed are a very recent biological adaptation. While people have the same reward pathway as rats—every mammal relies on the dopamine system — moral circuits can be found in only the most social primates. Humans, of course, are the most social primates of all.

The best way to probe the unique brain circuits underlying morality is by using a brain scanner to study people while they are making moral decisions. Consider this elegant experiment, led by neuroscientist Joshua Greene of Harvard. Greene asked his subjects a series of questions involving a runaway trolley, an oversize man, and five maintenance workers. (It might sound like a strange setup, but it's actually based on a well-known philosophical thought puzzle.) The first scenario goes like this:

> You are the driver of a runaway trolley. The brakes have failed. The trolley is approaching a fork in the track at top speed. If you do nothing, the train will stay left, where it will run over five maintenance workers who are fixing the track. All five workers will die. However, if you steer the train right — this involves flicking a switch and turning the wheel — you will swerve onto a track where there is one maintenance worker. What do you do? Are you willing to intervene and change the path of the trolley?

In this hypothetical case, about 95 percent of people agree that it is morally permissible to turn the trolley. The decision is just simple arithmetic: it's better to kill fewer people. Some moral philosophers even argue that it is immoral *not* to turn the trolley, since passivity will lead to the death of four more people. But what about this scenario:

> You are standing on a footbridge over the trolley track. You see a trolley racing out of control, speeding toward five workmen who are fixing the track. All five men will die unless the trolley can be stopped. Standing next to you on the footbridge is a very large man. He is leaning over the railing, watching the trolley hurtle toward the men. If you sneak up on the man and give him a little push, he will fall over the railing and into the path of the trolley. Because he is so big, he will stop the trolley

from killing the maintenance workers. Do you push the man off the footbridge? Or do you allow five men to die?

The brute facts, of course, remain the same: one man must die in order for five men to live. If ethical decisions were perfectly rational, then a person would act the same way in both situations and be as willing to push the man off the bridge as he or she was to turn the trolley. And yet, almost nobody is willing to actively throw another person onto the train tracks. The decisions lead to the same outcome, yet one is moral and one is murder.

Greene argues that pushing the man feels wrong because the killing is direct: you are using your body to hurt his body. He calls it a *personal* moral situation, since it directly involves another person. In contrast, when you just have to turn the trolley onto a different track, you aren't directly hurting somebody else, you're just shifting the trolley wheels; the resulting death seems indirect. In this case, it's an *impersonal* moral decision.

What makes this thought experiment so interesting is that the fuzzy moral distinction—the difference between personal and impersonal decisions—is built into the brain. It doesn't matter what culture you live in, or what religion you subscribe to: the two different trolley scenarios trigger distinct patterns of activation. In the first scenario, when a subject was asked whether the trolley should be turned, the rational decision-making machinery was turned on. A network of brain regions assessed the various alternatives, sent their verdict onward to the prefrontal cortex, and the person chose the clearly superior option. The brain quickly realized that it was better to kill one man than five men.

However, when a subject was asked whether he would be willing to push a man onto the tracks, a separate network of brain areas was activated. These folds of gray matter—the superior temporal sulcus, posterior cingulate, and medial frontal gyrus—are responsible for interpreting the thoughts and feelings of *other* people. As a result, the subject automatically imagined how the poor man would feel as he plunged to his death on the train tracks below. He vividly simulated his mind and concluded that pushing him was a capital crime, even if it saved the lives of five other men. The person couldn't explain the moral decision—the inner lawyer was confused by the inconsistency—but his certainty never wavered. Pushing a man off a bridge just *felt* wrong.

While stories of Darwinian evolution often stress the amorality of natural selection—we are all Hobbesian brutes, driven to survive by selfish genes—our psychological reality is much less bleak. We aren't angels, but we also aren't depraved hominids. "Our primate ancestors," Greene explains, "had intensely social lives. They evolved mental mechanisms to keep them from doing all the nasty things they might otherwise be interested in doing. This basic primate morality doesn't understand things like tax evasion, but it does understand things like pushing your buddy off of a cliff." As Greene puts it, a personal moral violation can be roughly defined as "me hurts you," a concept simple enough for a primate to understand.

This is a blasphemous idea. Religious believers assume that God invented the moral code. It was given to Moses on Mount Sinai, a list of imperatives inscribed in stone. (As Dostoyevsky put it, "If there is no God, then we are lost in a moral chaos. Everything is permitted.") But this cultural narrative gets the causality backward. Moral emotions existed long before Moses. They are writ into the primate brain. Religion simply allows us to codify these intuitions, to translate the ethics of evolution into a straightforward legal system. Just look at the Ten Commandments. After God makes a series of religious demands—don't worship idols and always keep the Sabbath—He starts to issue moral orders. The first order is the foundation of primate morality: thou shalt not kill. Then comes a short list of moral adjuncts, which are framed in terms of harm to another human being. God doesn't tell us merely not to lie; He tells us not to bear false witness against our neighbor. He doesn't prohibit jealousy only in the abstract; He commands us not to covet our neighbor's "wife or slaves or ox or donkey." The God of the Old Testament understands that our most powerful moral emotions are generated in response to personal moral scenarios, so that's how He frames all of His instructions. The details of the Ten Commandments reflect the details of the evolved moral brain.

These innate emotions are so powerful that they keep people moral even in the most amoral situations. Consider the behavior of soldiers during war. On the battlefield, men are explicitly encouraged to kill one another; the crime of murder is turned into an act of heroism. And yet, even in such violent situations, soldiers often struggle to get past their moral instincts. During World War II, for example, U.S. Army Brigadier General S.L.A. Marshall undertook a survey of thousands of American troops right after they'd been in combat. His shocking conclusion was that less than 20 percent actually shot at the enemy, even when under attack. "It is fear of killing," Marshall wrote, "rather than fear of being killed, that is the most common cause of battle failure in the individual." When soldiers were forced to confront the possibility of directly harming other human beings—this is a personal moral decision—they were literally incapacitated by their emotions. "At the most vital point of battle," Marshall wrote, "the soldier becomes a conscientious objector."

After these findings were published, in 1947, the U.S. Army realized it had a serious problem. It immediately began revamping its training regimen in order to increase the "ratio of fire." New recruits began endlessly rehearsing the kill, firing at anatomically correct targets that dropped backward after being hit. As Lieutenant Colonel Dave Grossman noted, "What is being taught in this environment is the ability to shoot reflexively and instantly . . . Soldiers are de-sensitized to the act of killing, until it becomes an automatic response." The army also began emphasizing battlefield tactics, such as high-altitude bombing and long-range artillery, that managed to obscure the personal cost of war. When bombs are dropped from forty thousand feet, the decision to fire is like turning a trolley wheel: people are detached from the resulting deaths.

These new training techniques and tactics had dramatic results. Several years after he published his study, Marshall was sent to fight in the Korean War, and he discovered that 55 percent of infantrymen were now firing their weapons. In Vietnam, the ratio of fire was nearly 90 percent. The army had managed to turn the most personal of moral situations into an impersonal reflex. Soldiers no longer felt a surge of negative emotions when they fired their weapons. They had been turned, wrote Grossman, into "killing machines."

3

At its core, moral decision-making is about sympathy. We abhor violence because we know violence hurts. We treat others fairly because we know what it feels like to be treated unfairly. We reject suffering because we can imagine what it's like to suffer. Our minds naturally bind us together, so that we can't help but follow the advice of Luke: "And as ye would that men should do to you, do ye also to them likewise."

Feeling sympathetic is not as simple as it might seem. For starters, before you can sympathize with the feelings of other people, you have to figure out what they are feeling. This means you need to develop a theory about what's happening inside their minds so that your emotional brain can imitate the activity of their emotional brains. Sometimes, this act of mind reading is done by interpreting facial expressions. If someone is squinting his eyes and clenching his jaw, you automatically conclude that

his amygdala is excited; he must be angry. If he flexes the zygo-
maticus majors — that's what happens during a smile — then you
assume he's happy. Of course, you don't always have access to a
communicative set of facial expressions. When you talk on the
phone or write an e-mail or think about someone far away, you
are forced to mind read by simulation, by imagining what you
would feel in the same situation.

Regardless of how exactly one generates theories of other
people's minds, it's clear that these theories profoundly affect
moral decisions. Look, for example, at the ultimatum game, a
staple of experimental economics. The rules of the game are simple, if a little bit unfair: an experimenter pairs two people together, and hands one of them ten dollars. This person (the proposer) gets to decide how the ten dollars is divided. The second
person (the responder) can either accept the offer, which allows
both players to pocket their respective shares, or reject the offer,
in which case both players walk away empty-handed.

When economists first started playing this game in the early
1980s, they assumed that this elementary exchange would always generate the same outcome. The proposer would offer the
responder about a dollar — a minimal amount — and the responder would accept it. After all, a rejection leaves both players
worse off, and one dollar is better than nothing, so this arrange-
ment would clearly demonstrate our innate selfishness and rationality.

However, the researchers soon realized that their predictions
were all wrong. Instead of swallowing their pride and pocketing
a small profit, responders typically rejected any offer they perceived as unfair. Furthermore, proposers anticipated this angry
rejection and typically tendered an offer of around five dollars.
This was such a stunning result that nobody really believed it.

But when other scientists repeated the experiment, the same
thing happened. People play this game the same way all over the
world, and studies have observed similar patterns of irrationality

in Japan, Russia, Germany, France, and Indonesia. No matter where the game was played, people almost always made fair offers. As the economist Robert Frank notes, "Seen through the lens of modern self-interest theory, such behavior is the human equivalent of planets traveling in square orbits."

Why do proposers engage in such generosity? The answer returns us to the act of sympathy and the unique brain circuits that determine moral decisions. Adam Smith, the eighteenth-century philosopher, was there first. Although Smith is best known for his economic treatise *The Wealth of Nations,* he was most proud of *The Theory of Moral Sentiments,* his sprawling investigation into the psychology of morality. Like his friend David Hume, Smith was convinced that our moral decisions were shaped by our emotional instincts. People were good for essentially irrational reasons.

According to Smith, the source of these moral emotions was the imagination, which people used to mirror the minds of others. (The reflective mirror, which had recently become a popular household item in Smith's time, is an important metaphor in his writing on morality.) "As we have no immediate experience of what other men feel," Smith wrote, "we can form no idea of the manner in which they are affected, but by conceiving what we ourselves should feel in the like situation." This mirroring process leads to an instinctive sympathy for one's fellow man — Smith called it "fellow-feeling" — that forms the basis for moral decisions.

Smith was right. The reason a proposer makes a fair offer in the ultimatum game is that he is able to imagine how the responder will feel about an unfair offer. (When people play the game with computers, they are never generous.) The responder knows that a low-ball proposal will make the other person angry, which will lead him to reject the offer, which will leave everybody with nothing. So the proposer suppresses his greed and equitably splits the ten dollars. That ability to sympathize with the feelings of others leads to fairness.

The sympathetic instinct is also one of the central motivations behind altruism, which is demonstrated when people engage in selfless acts such as donating to charity and helping out perfect strangers. In a recent experiment published in *Nature Neuroscience,* scientists at Duke University imaged the brains of people as they watched a computer play a simple video game. Because the subjects were told that the computer was playing the game for a specific purpose — it wanted to earn money — their brains automatically treated the computer like an "intentional agent," complete with goals and feelings. (Human minds are so eager to detect other minds that they often imbue inanimate objects, like computers and stuffed animals, with internal mental states.) Once that happened, the scientists were able to detect activity in the superior temporal sulcus and other specialized areas that help each of us theorize and sympathize with the emotions of other people. Even though the subjects knew they were watching a computer, they couldn't help but imagine what the computer was feeling.

Now comes the interesting part: the scientists noticed that there was a lot of individual variation during the experiment. Some people had very active sympathetic brains, while others seemed rather uninterested in thinking about the feelings of someone else. The scientists then conducted a survey of altruistic behavior, asking people how likely they would be to "help a stranger carry a heavy object" or "let a friend borrow a car." That's when the correlation became clear: people who showed more brain activity in their sympathetic regions were also much more likely to exhibit altruistic behavior. Because they intensely imagined the feelings of other people, they wanted to make other people feel better, even if it came at personal expense.

But here's the lovely secret of altruism: *it feels good.* The brain is designed so that acts of charity are pleasurable; being nice to others makes us feel nice. In a recent brain-imaging experiment, a few dozen people were each given $128 of real money and allowed to choose between keeping the money and donating it to charity. When they chose to give away the money, the reward centers of their brains became active and they experienced the delightful glow of unselfishness. In fact, several subjects showed more reward-related brain activity during acts of altruism than they did when they actually received cash rewards. From the perspective of the brain, it literally was better to give than to receive.

ONE OF THE ways neuroscientists learn about the brain is by studying what happens when something goes wrong with it. For example, scientists learned about the importance of our moral emotions by studying psychopaths; they learned about the crucial role of dopamine by studying people with Parkinson's; and brain tumors in the frontal lobes have helped to illuminate the substrate of rationality. This might seem callous—tragedy is turned into an investigative tool—but it's also extremely effective. The broken mind helps us understand how the normal mind works.

When it comes to the sympathetic circuits in the human brain, scientists have learned a tremendous amount by studying people with autism. When Dr. Leo Kanner first diagnosed a group of eleven children with autism, in 1943, he described the syndrome as one of "extreme aloneness." (*Aut* is Greek for "self," and *autism* translates to "the state of being unto one's self.") The syndrome afflicts one in every 160 individuals and leaves them emotionally isolated, incapable of engaging in many of the social interactions most people take for granted. As the Cambridge psychologist Simon Baron-Cohen puts it, people with autism are "mind-blind." They have tremendous difficulty interpreting the emotions and mental states of others.*

Scientists have long suspected that autism is a disease of brain development. For some still mysterious reason, the cortex doesn't wire itself correctly during the first year of life. It now appears that one of the brain areas compromised in people with autism is a small cluster of cells known as mirror neurons. The name of the cell type is literal: these neurons mirror the movements of other people. If you see someone else smile, then your mirror neurons will light up as if you were smiling. The same thing happens whenever you see someone scowl, grimace, or cry. These cells reflect, on the inside, the expressions of everybody else. As Giacomo Rizzolatti, one of the scientists who discovered mirror neurons, says, "They [mirror neurons] allow us to grasp the minds of others not through conceptual reasoning but through direct simulation; by feeling, not by thinking."

This is what people with autism have to struggle to do. When scientists at UCLA imaged the brains of autistic people as they looked at photographs of faces in different emotional states, the scientists discovered that the autistic brain, unlike the normal brain, showed no activity in its mirror-neuron area. As a result, the autistic subjects had difficulty interpreting the feelings on display. They saw the angry face as nothing but a set of flexed facial muscles. A happy face was simply a different set of muscles. But neither expression was correlated with a specific emotional state. In other words, they never developed a theory about what was happening inside other people's minds.

---

*Autism, obviously, has nothing to do with psychopathy. Unlike people with autism, psychopaths can readily recognize when others are upset or in pain. Their problem is that they can't generate corresponding emotions, since their amygdalas are never turned on. The end result is that psychopaths remain preternaturally calm, even in situations that should make them upset. People with autism, however, don't have a problem generating emotion. The problem for them is of recognition; they struggle to decipher or simulate the mental states of others.

A brain-imaging study done by scientists at Yale sheds further light on the anatomical source of autism. The study examined the parts of the brain that were activated when a person looked at a face and when he or she looked at a static object, like a kitchen chair. Normally, the brain reacts very differently to these stimuli. Whenever you see a human face, you use a highly specialized brain region called the fusiform face area (FFA) that is solely devoted to helping you recognize other people. In contrast, when you look at a chair, the brain relies on the inferior temporal gyrus, an area activated by any sort of complex visual scene. However, in the study, people with autism never turned on the fusiform face area. They looked at human faces with the part of the brain that normally recognizes objects. A person was just another thing. A face generated no more emotion than a chair.

These two brain deficits—a silent mirror-neuron circuit and an inactive fusiform face area—help to explain the social difficulties of people with autism. Their "extreme aloneness" is a direct result of not being able to interpret and internalize the emotions of other people. Because of this, they often make decisions that, in the words of one autism researcher, "are so rational they can be hard to understand."

For instance, when people with autism play the ultimatum game, they act just like the hypothetical agents in an economics textbook. They try to apply a rational calculus to the irrational world of human interaction. On average, they make offers that are 80 percent lower than those of normal subjects, with many offering less than a nickel. This greedy strategy ends up being ineffective, since the angry responders tend to reject such unfair offers. But the proposers with autism are unable to anticipate these feelings. Consider this quote from an upset autistic adult whose offer of ten cents in a ten-dollar ultimatum game was spurned: "I did not earn any money because all the other players are *stupid*! How can you reject a positive amount of money and prefer to get zero? They just did not understand the game!

You should have stopped the experiment and explained it to them . . ."

Autism is a chronic condition, a permanent form of mind blindness. But it's possible to induce a temporary state of mind blindness, in which the brain areas that normally help a person sympathize with others are turned off. A simple variation on the ultimatum game, known as the dictator game, makes this clear. Our sense of "fellow-feeling" is natural, but it's also very fragile. Unlike the ultimatum game, in which the responder can decide whether or not to accept the monetary offer, in the dictator game, the proposer simply dictates how much the responder receives. What's surprising is that these tyrants are still rather generous and give away about one-third of the total amount of money. Even when people have absolute power, they remain constrained by their sympathetic instincts.

However, it takes only one minor alteration for this benevolence to disappear. When the dictator cannot see the responder —the two players are located in separate rooms—the dictator lapses into unfettered greed. Instead of giving away a significant share of the profits, the despots start offering mere pennies and pocketing the rest. Once people become socially isolated, they stop simulating the feelings of other people. Their moral intuitions are never turned on. As a result, the inner Machiavelli takes over, and the sense of sympathy is squashed by selfishness. The UC Berkeley psychologist Dacher Keltner has found that in many social situations, people with power act just like patients with damage to the emotional brain. "The experience of power might be thought of as having someone open up your skull and take out that part of your brain so critical to empathy and socially appropriate behavior," he says. "You become very impulsive and insensitive, which is a bad combination."

Paul Slovic, a psychologist at the University of Oregon, has exposed another blind spot in the sympathetic brain. His experiments are simple: he asks people how much they would be willing to donate to various charitable causes. For example, Slovic found that when people were shown a picture of Rokia, a starving Malawian child, they acted with impressive generosity. After looking at Rokia's emaciated body and haunting brown eyes, they each donated, on average, two dollars and fifty cents to the charity Save the Children. However, when other people were provided with a list of statistics about starvation throughout Africa—more than three million children in Malawi are malnourished, more than eleven million people in Ethiopia need immediate food assistance, and so forth—the average donation was 50 percent lower. At first glance, this makes no sense. When people are informed about the true scope of the problem, they should give *more* money, not less. Rokia's tragic story is just the tip of the iceberg.

According to Slovic, the problem with statistics is that they don't activate our moral emotions. The depressing numbers leave us cold: our minds can't comprehend suffering on such a massive scale. This is why we are riveted when one child falls down a well but turn a blind eye to the millions of people who die every year for lack of clean water. And why we donate thousands of dollars to help a single African war orphan featured on the cover of a magazine but ignore widespread genocides in Rwanda and Darfur. As Mother Teresa put it, "If I look at the mass, I will never act. If I look at the one, I will."

4

The capacity for making moral decisions is innate—the sympathetic circuit is hard-wired, at least in most of us—but it still requires the right kind of experience in order to develop. When everything goes according to plan, the human mind naturally develops a potent set of sympathetic instincts. We will resist pushing the man off the bridge, make fair offers in the ultimatum game, and get deeply disturbed by images of other people in pain.

However, if something goes amiss during the developmental process—if the circuits that underlie moral decisions never mature—the effects can be profound. Sometimes, as with autism, the problem is largely genetic. (Scientists estimate the heritability of autism at somewhere between 80 and 90 percent, which makes it one of the most inheritable of all neurologic conditions.) But there's another way that the developing brain can be permanently damaged: child abuse. When children are molested or neglected or unloved as children, their emotional brains are warped. (John Gacy, for example, was physically abused throughout his childhood by his alcoholic father.) The biological program that allows human beings to sympathize with the feelings of others is turned off. Cruelty makes us cruel. Abuse makes us abusive. It's a tragic loop.

The first evidence for this idea came from the work of Harry Harlow.* In the early 1950s, Harlow decided to start a breeding colony of monkeys at the University of Wisconsin. He was studying Pavlovian conditioning in primates, but he needed more data, which meant that he needed more animals. Although nobody had ever successfully bred monkeys in the United States before then, Harlow was determined.

The breeding colony began with just a few pregnant female monkeys. Harlow carefully monitored the expecting primates; after each gave birth, he immediately isolated the infant in an immaculately clean cage. At first, everything went according to plan. Harlow raised the babies on a formula of sugar and evaporated milk fortified with a slew of vitamins and supplements. He fed the monkeys from sterilized doll bottles every two hours and carefully regulated the cycles of light and dark. In order to mini-

---

*For a wonderful history of Harlow and his research, see Deborah Blum's biography *Love at Goon Park.*

mize the spread of disease, Harlow never let the babies interact with one another. The result was a litter of primates that were bigger and stronger than their peers from the wild.

But the physical health of these young monkeys hid a devastating sickness: they had been wrecked by loneliness. Their short lives had been defined by total isolation, and they proved incapable of even the most basic social interactions. They would maniacally rock back and forth in their metal cages, sucking on their thumbs until they bled. When they encountered other monkeys, they would shriek in fear, run to the corners of their cages, and stare at the floor. If they felt threatened, they would lash out in vicious acts of violence. Sometimes these violent tendencies were turned inward. One monkey ripped out its fur in bloody clumps. Another gnawed off its own hand. Because of their early deprivation, these babies had to be isolated for the rest of their lives.

For Harlow, these troubled baby monkeys demonstrated that the developing mind needed more than proper nutrition. But what did it need? The first clue came from watching these primate babies. The scientists had lined their cages with cloth diapers so that the monkeys didn't have to sleep on the cold concrete floor. The motherless babies quickly became obsessed with these cloth rags. They would wrap themselves in the fabric and cling to the diapers if anybody approached the cages. The soft fabric was their sole comfort.

This poignant behavior inspired Harlow to come up with a new experiment. He decided to raise the next generation of baby monkeys with two different pretend mothers. One was a wire mother, formed out of wire mesh, while the other was a mother made out of soft terry cloth. Harlow assumed that all things being equal, the babies would prefer the cloth mothers, since they would be able to cuddle with the fabric. To make the experiment more interesting, Harlow added a slight twist to a few of the cages. Instead of hand-feeding some babies, he put their milk bottles in the hands of the wire mothers. His question was simple: what was more important, food or affection? Which mother would the babies want more?

In the end, it wasn't even close. No matter which mother held the milk, the babies always preferred the cloth mothers. The monkeys would run over to the wire mothers and quickly sate their hunger before immediately returning to the comforting folds of cloth. By the age of six months, the babies were spending more than eighteen hours a day nuzzling with their soft parent. They were with the wire mothers only long enough to eat.

The moral of Harlow's experiment is that primate babies are born with an intense need for attachment. They cuddled with the cloth mothers because they wanted to experience the warmth and tenderness of a real mother. Even more than food, these baby monkeys craved the feeling of affection. "It's as if the animals are programmed to seek out love," Harlow wrote.

When this need for love wasn't met, the babies suffered from a tragic list of side effects. The brain was permanently damaged so that the monkeys with wire mothers didn't know how to deal with others, sympathize with strangers, or behave in a socially acceptable manner. Even the most basic moral decisions were impossible. As Harlow would later write, "If monkeys have taught us anything, it's that you've got to learn how to love before you learn how to live."

Harlow would later test the limits of animal experimentation, remorselessly probing the devastating effects of social isolation. His cruelest experiment was putting baby monkeys in individual cages with nothing—not even a wire mother—for months at a time. The outcome was unspeakably sad. The isolated babies were like primate psychopaths, completely numb to all expressions of emotion. They started fights without provocation and they didn't stop fighting until one of the monkeys had been seriously injured. They were even vicious to their own children. One psychopathic monkey bit off the fingers of her child. Another killed her crying baby by crushing its head in her mouth. Most psychopathic mothers, however, just perpetuated the devastating cycle of cruelty. When their babies tried to cuddle, they would push them away. The confused infants would try again and again, but to no avail. Their mothers felt nothing.

WHAT HAPPENS TO monkeys can happen to people. This is the tragic lesson of Communist Romania. In 1966, Nicolae Ceausescu, the despotic leader of the country, banned all forms of contraception, and the country was suddenly awash in unwanted babies. The predictable result was a surfeit of orphans; poor families surrendered the kids they couldn't afford.

The state-run orphanages of Romania were overwhelmed and underfunded. Babies were left in cribs with nothing but plastic bottles. Toddlers were tied to their beds and never touched. The orphanages lacked heat in the winter. Children with disabilities were consigned to the basement, and some went years without seeing natural light. Older children were drugged so that they would sleep for days at a time. In some orphanages, more than 25 percent of the children died before the age of five.

The children that managed to survive the Romanian orphanages were permanently scarred. Many had stunted bodies, shrunken bones, and untreated infections. But the most devastating legacy of the orphanage system was psychological. Many of the abandoned children suffered from severe emotional impairments. They were often hostile to strangers, abusive to one another, and incapable of even the most basic social interactions. Couples that adopted Romanian orphans from these institutions reported a wide array of behavioral disorders. Some children cried whenever they were touched. Others stared into space for hours and then suddenly flew into violent rages, attacking everything within reach. One Canadian couple walked into the bedroom of their three-year-old son to discover that he had just thrown their new kitten out the window.

When neuroscientists imaged the brain activity of Romanian orphans, they saw reduced activity in regions that are essential for emotion and social interaction, such as the orbitofrontal cortex and the amygdala. The orphans also proved unable to perceive emotions in others and had a pronounced inability to interpret facial expressions. Finally, the neglected children showed significantly reduced levels of vasopressin and oxytocin, two hormones crucial for the development of social attachments. (These hormonal deficiencies persisted for years afterward.) For these victims of abuse, the world of human sympathy was incomprehensible. They struggled to recognize the emotions of others, and they also found it difficult to modulate their own emotions.

Studies of American children who are abused at an early age paint a similarly bleak picture. In the early 1980s, the psychologists Mary Main and Carol George looked at a group of twenty toddlers from "families in stress." Half of these children had been victims of serious physical abuse. The other half were from broken homes — many of them were living with foster parents — but they hadn't been hit or hurt. Main and George wanted to see how these two groups of disadvantaged toddlers responded to a crying classmate. Would they display normal human sympathy? Or would they be unable to relate to the feelings of their peer? The researchers found that almost all the nonabused children reacted to the upset child with concern. Their instinctive sympathy led them to make some attempts to console the child. They were upset by seeing somebody else upset.

Childhood abuse, however, changed everything. The abused toddlers didn't know how to react to their distressed classmate. They occasionally made sympathetic gestures, but these gestures often degenerated into a set of aggressive threats if the other child didn't stop crying. Here is the study's description of Martin, an abused two-and-a-half-year-old: "Martin . . . tried to take the hand of the crying other child, and when she resisted, he slapped her on the arm with his open hand. He then turned away from her to look at the ground and began vocalizing very strongly. 'Cut it out! cut it out!,' each time saying it a little faster and louder. He patted her, but when she became disturbed by his patting, he retreated, hissing at her and baring his teeth. He then began patting her on the back again, his patting became beating, and he continued beating her despite her screams." Even when Martin wanted to help, he ended up making things worse. An abused two-year-old named Kate exhibited a similar pattern of behavior. At first she reacted with tenderness to the distressed child and gently caressed him on the back. "Her patting, however, soon became very rough," the researchers wrote, "and she began hitting him hard. She continued to hit him until he crawled away." Because Kate and Martin couldn't understand the feelings of someone else, the world of human interaction had become an impenetrable place.

What these abused children were missing was an education in feeling. Because they had been denied that influx of tender emotion that the brain is built to expect, they were seriously scarred, at least on the inside. It's not that these kids wanted to be cruel or unsympathetic. They were simply missing the patterns of brain activity that normally guide our moral decisions. As a result, they reacted to the toddler in distress just as their abusive parents reacted to their own distress: with threats and violence.

But these tragic examples are exceptions to the rule. We are designed to feel one another's pain so that we're extremely distressed when we hurt others and commit moral transgressions. Sympathy is one of humanity's most basic instincts, which is why evolution lavished so much attention on mirror neurons, the fusiform face area, and those other brain regions that help theorize about other minds. As long as a person is loved as a child and doesn't suffer from any developmental disorders, the human brain will naturally reject violence and make fair offers and try to comfort the crying child. This behavior is just a basic part of who we are. Evolution has programmed us to care about one another.

Consider this poignant experiment: six rhesus monkeys were trained to pull on a variety of chains to get food. If they pulled on one chain, they got a large amount of their favorite food. If they pulled on a different chain, they got a small amount of a less enticing food. As you can probably guess, the monkeys quickly learned to pull on the chain that gave them more of what they wanted. They maximized their reward.

After a few weeks of this happy setup, one of the six monkeys got hungry and decided to pull on the chain. This is when something terrible happened: a separate monkey in a different cage was shocked with a painful jolt of electricity. All six monkeys saw it happen. They heard the awful shriek. They watched the monkey grimace and cower in fear. The change in their behavior was immediate. Four of the monkeys decided to stop pulling on the maximizing chain. They were now willing to settle for less food as long as the other monkey wasn't hurt. The fifth monkey stopped pulling on either chain for five days, and the sixth monkey stopped pulling for twelve days. They starved themselves so that a monkey they didn't know wasn't forced to suffer.

# 7

# The Brain Is an Argument

One of the most coveted prizes in a presidential primary is the endorsement of the *Concord Monitor,* a small newspaper in central New Hampshire. During the first months of the 2008 presidential primary campaign, all of the major candidates, from Chris Dodd to Mike Huckabee, sat for interviews with the paper's editorial board. Some candidates, such as Hillary Clinton, Barack Obama, and John McCain, were invited back for follow-up interviews. These sessions would often last for hours, with the politicians facing a barrage of uncomfortable questions. Hillary Clinton was asked about various White House scandals; Barack Obama was asked why he often seemed "bored and low-key" on the stump; McCain was asked about his medical history. "There were a few awkward moments," says Ralph Jimenez, the editorial-page editor. "You could tell they were thinking, *Did you just ask me that? Do you know who I am?*"

But the process wasn't limited to these interviews. Bill Clinton got in the habit of calling the editors, at home and on their cell phones, and launching into impassioned defenses of his wife.

(Some of the editors had unlisted phone numbers, which made Clinton's calls even more impressive.) Obama had his own persistent advocates. The board was visited by former White House staff members, such as Madeleine Albright and Ted Sorensen, and lobbied by a bevy of local elected officials. For the five members of the editorial board, all the attention was flattering, if occasionally annoying. Felice Belman, the executive editor of the *Monitor,* was awakened by a surprise phone call from Hillary at seven thirty on a Saturday morning. "I was still half asleep," she says. "And I definitely wasn't in the mood to talk about health-care mandates." (Ralph still has a phone message from Hillary Clinton on his cell phone.)

Twelve days before the primary, on a snowy Thursday afternoon, the editorial board gathered in a back office of the newsroom. They'd postponed the endorsement meeting long enough; it was time to make a decision. Things would be easy on the Republican side: all five members favored John McCain. The Democrat endorsement, however, was a different story. Although the editors had each tried to keep an open mind — "The candidates are here for a year and you don't want to settle on one candidate right away," said Mike Pride, a former editor of the paper — the room was starkly divided into two distinct camps. Ralph Jimenez and Ari Richter, the managing editor, were pushing for an Obama endorsement. Mike Pride and Geordie Wilson, the publisher, favored Clinton. And then there was Felice, the sole undecided vote. "I was waiting to be convinced until the last minute," she says. "I guess I was leaning toward Clinton, but I still felt like I could have been talked into switching sides."

Now came the hard part. The board began by talking about the issues, but there wasn't that much to talk about: Obama and Clinton had virtually identical policy positions. Both candidates were in favor of universal health care, repealing the Bush tax cuts, and withdrawing troops from Iraq as soon as possible. And yet, despite this broad level of agreement, the editors were fiercely loyal to their chosen candidates, even if they couldn't explain *why* they were so loyal. "You just know who you prefer," Ralph says. "For most of the meeting, the level of discourse was pretty much 'My person is better. Period. End of story.'"

After a lengthy and intense discussion—"We'd really been having this discussion for months," says Ralph—the *Monitor* ended up endorsing Clinton by a 3–2 vote. The room was narrowly split, but it had become clear that no one was going to change his or her mind. Even Felice, the most uncertain of the editors, was now firmly in the Clinton camp. "There is always going to be disagreement," Mike says. "That's what happens when you get five opinionated people in the same room talking politics. But you also know that before you leave the room, you've got to endorse somebody. You've got to accept the fact that some people are bound to be wrong"—he jokingly looks over at Ralph—"and find a way to make a decision."

For readers of the *Monitor,* the commentary endorsing Clinton seemed like a well-reasoned brief, an unambiguous summary of the newspaper's position. (Kathleen Strand, the Clinton spokesperson in New Hampshire, credited the endorsement with helping Clinton win the primary.) The carefully chosen words in the editorial showed no trace of the debate that had plagued the closed-door meeting and all those heated conversations by the water cooler. If just one of the editors had changed his or her mind, then the *Monitor* would have chosen Obama. In other words, the clear-cut endorsement emerged from a very tentative majority.

In this sense, the editorial board is a metaphor for the brain. Its decisions often feel unanimous—you know which candidate you prefer—but the conclusions are actually reached only after a series of sharp internal disagreements. While the cortex struggles to make a decision, rival bits of tissue are contradicting one another. Different brain areas think different things for different reasons. Sometimes this fierce argument is largely emotional, and the distinct parts of the limbic system debate one another. Although people can't always rationally justify their feelings—these editorial board members preferred either Hillary or Obama for reasons they couldn't really articulate—these feelings still manage to powerfully affect behavior. Other arguments unfold largely between the emotional and rational systems of the brain as the prefrontal cortex tries to resist the impulses coming from below. Regardless of which areas are doing the arguing, however, it's clear that all those mental components stuffed inside the head are constantly fighting for influence and attention. Like an editorial board, the mind is an extended argument. And it is arguing with itself.

In recent years, scientists have been able to show that this "argument" isn't confined only to contentious issues such as presidential politics. Rather, it's a defining feature of the decision-making process. Even the most mundane choices emerge from a vigorous cortical debate. Let's say, for instance, that you're contemplating breakfast cereals in the supermarket. Each option will activate a unique subset of competing thoughts. Perhaps the organic granola is delicious but too expensive, or the whole-grain flakes are healthy but too unappetizing, or the Fruit Loops are an appealing brand (the advertisements worked) but too sugary. Each of these distinct claims will trigger a particular set of emotions and associations, all of which then compete for your conscious attention. Antoine Bechara, a neuroscientist at USC, compares this frantic neural competition to natural selection, with the stronger emotions ("I really want Honey Nut Cheerios!") and the more compelling thoughts ("I should eat more fiber") gaining a selective advantage over weaker ones ("I like the cartoon character on the box of Fruit Loops"). "The point is that most of the computation is done at an emotional, unconscious level, and not at a logical level," he says. The particular ensemble of brain cells that win the argument determine what you eat for breakfast.

Consider this clever experiment designed by Brian Knutson and George Loewenstein. The scientists wanted to investigate what happens inside the brain when a person makes typical consumer choices, such as buying an item in a retail store or choosing a cereal. A few dozen lucky undergraduates were recruited as experimental subjects and given a generous amount of spending money. Each subject was then offered the chance to buy dozens of different objects, from a digital voice recorder to gourmet chocolates to the latest Harry Potter book. After the student stared at each object for a few seconds, he was shown the price tag. If he chose to buy the item, its cost was deducted from the original pile of cash. The experiment was designed to realistically simulate the experience of a shopper.

While the student was deciding whether or not to buy the product on display, the scientists were imaging the subject's brain activity. They discovered that when a subject was first exposed to an object, his nucleus accumbens (NAcc) was turned on. The NAcc is a crucial part of the dopamine reward pathway, and the intensity of its activation was a reflection of desire for the item. If the person already owned the complete Harry Potter collection, then the NAcc didn't get too excited about the prospect of buying another copy. However, if he had been craving a George Foreman grill, the NAcc flooded the brain with dopamine when that item appeared.

But then came the price tag. When the experimental subject was exposed to the cost of the product, the insula and prefrontal cortex were activated. The insula produces aversive feelings and is triggered by things like nicotine withdrawal and pictures of people in pain. In general, we try to avoid anything that makes our insulas excited. This includes spending money. The prefrontal cortex was activated, scientists speculated, because this rational area was computing the numbers, trying to figure out if the product was actually a good deal. The prefrontal cortex got most excited during the experiment when the cost of the item on display was significantly lower than normal.

By measuring the relative amount of activity in each brain region, the scientists could accurately predict the subjects' shopping decisions. They knew which products people would buy before the people themselves did. If the insula's negativity exceeded the positive feelings generated by the NAcc, then the subject always chose not to buy the item. However, if the NAcc was more active than the insula, or if the prefrontal cortex was convinced that it had found a good deal, the object proved irresistible. The sting of spending money couldn't compete with the thrill of getting something new.

This data, of course, directly contradicts the rational models of microeconomics; consumers aren't always driven by careful considerations of price and expected utility. You don't look at the electric grill or box of chocolates and perform an explicit cost-benefit analysis. Instead, you outsource much of this calculation to your emotional brain and then rely on relative amounts of pleasure versus pain to tell you what to purchase. (During many of the decisions, the rational prefrontal cortex was largely a spectator, standing silently by while the NAcc and insula argued with each other.) Whichever emotion you feel most intensely tends to dictate your shopping decisions. It's like an emotional tug of war.

This research explains why consciously analyzing purchasing decisions can be so misleading. When Timothy Wilson asked people to analyze their strawberry-jam preferences, they made worse decisions because they had no idea what their NAccs really wanted. Instead of listening to their feelings, they tried to deliberately decipher their pleasure. But we can't ask our NAccs questions; we can only listen to what they have to say. Our desires exist behind locked doors.

Retail stores manipulate this cortical setup. They are designed to get us to open our wallets; the frivolous details of the shopping experience are really subtle acts of psychological manipulation. The store is tweaking our brains, trying to soothe the insulas and stoke the NAccs. Just look at the interior of a Costco warehouse. It's no accident that the most coveted items are put in the most prominent places. A row of high-definition televisions lines the entrance. The fancy jewelry, Rolex watches, iPods, and other luxury items are conspicuously placed along the corridors with the heaviest foot traffic. And then there are the free samples of food, liberally distributed throughout the store. The goal of Costco is to constantly prime the pleasure centers of the brain, to keep us lusting after things we don't need. Even though you probably won't buy the Rolex, just looking at the fancy watch makes you more likely to buy something else, since the desired item activates the NAcc. You have been conditioned to crave a reward.

But exciting the NAcc is not enough; retailers must also inhibit the insula. This brain area is responsible for making sure you don't get ripped off, and when it's repeatedly assured by retail stores that low prices are "guaranteed," or that a certain item is on sale, or that it's getting the "wholesale price," the insula stops worrying so much about the price tag. In fact, researchers have found that when a store puts a promotional sticker next to the price tag—something like "Bargain Buy!" or "Hot Deal!"—but doesn't actually reduce the price, sales of that item still dramatically increase. These retail tactics lull the brain into buying more things, since the insula is pacified. We go broke convinced that we are saving money.

This model of the shopping brain also helps explain why credit cards make us spend so irresponsibly. According to Knutson and Loewenstein, paying with plastic literally inhibits the insula, making a person less sensitive to the cost of an item. As a result, the activity of the NAcc—the pleasure pump of the cortex—becomes disproportionately important: it wins every shopping argument.

# 1

There's something unsettling about seeing the brain as one big argument. We like to believe that our decisions reflect a clear cortical consensus, that the entire mind agrees on what we should do. And yet, that serene self-image has little basis in reality. The NAcc might want the George Foreman grill, but the insula knows that you can't afford it, or the prefrontal cortex realizes that it's a bad deal. The amygdala might like Hillary Clinton's tough talk on foreign policy, but the ventral striatum is excited by Obama's uplifting rhetoric. These antagonistic reactions manifest themselves as a twinge of uncertainty. You don't know what you believe. And you certainly don't know what to do.

The dilemma, of course, is how to reconcile the argument. If the brain is always disagreeing with itself, then how can a person ever make a decision? At first glance, the answer seems obvious: force a settlement. The rational parts of the mind should intervene and put an end to all the emotional bickering.

While such a top-down solution might seem like a good idea —using the most evolutionarily advanced parts of the brain to end the cognitive contretemps—this approach must be used with great caution. The problem is that the urge to end the debate often leads to neglect of crucial pieces of information. A person is so eager to silence the amygdala, or quiet the OFC, or suppress some bit of the limbic system that he or she ends up making a bad decision. A brain that's intolerant of uncertainty—that can't stand the argument—often tricks itself into thinking the wrong thing. What Mike Pride says about editorial boards is also true of the cortex: "The most important thing is that everyone has their say, that you listen to the other side and try to understand their point of view. You can't short-circuit the process."

Unfortunately, the mind often surrenders to the temptation of shoddy top-down thinking. Just look at politics. Voters with strong partisan affiliations are a case study in how *not* to form opinions: their brains are stubborn and impermeable, since they already know what they believe. No amount of persuasion or new information is going to change the outcome of their mental debates. For instance, an analysis of five hundred voters with "strong party allegiances" during the 1976 campaign found that during the heated last two months of the contest, only sixteen people were persuaded to vote for the other party. Another study tracked voters from 1965 to 1982, tracing the flux of party affiliation over time. Although it was an extremely tumultuous era in American politics — there was the Vietnam War, stagflation, the fall of Richard Nixon, oil shortages, and Jimmy Carter — nearly 90 percent of people who identified themselves as Republicans in 1965 ended up voting for Ronald Reagan in 1980. The happenings of history didn't change many minds.

It's now possible to see why partisan identities are so persistent. Drew Westen, a psychologist at Emory University, imaged the brains of ordinary voters with strong party allegiances during the run-up to the 2004 election. He showed the voters multiple, clearly contradictory statements made by each candidate, John Kerry and George Bush. For example, the experimental subject would read a quote from Bush praising the service of soldiers in the Iraq war and pledging "to provide the best care for all veterans." Then the subject would learn that on the same day Bush made this speech, his administration cut medical benefits for 164,000 veterans. Kerry, meanwhile, was quoted making contradictory statements about his vote to authorize war in Iraq.

After being exposed to the political inconsistencies of both candidates, the subject was asked to rate the level of contradiction on a scale of 1 to 4, with 4 signaling a strong level of contradiction. Not surprisingly, the reactions of voters were largely determined by their partisan allegiances. Democrats were troubled by Bush's inconsistent statements (they typically rated them a 4) but found Kerry's contradictions much less worrisome. Republicans responded in a similar manner; they excused Bush's gaffes but almost always found Kerry's statements flagrantly incoherent.

By studying each of these voters in an fMRI machine, Westen was able to look at the partisan reasoning process from the perspective of the brain. He could watch as Democrats and Republicans struggled to maintain their political opinions in the face of conflicting evidence. After being exposed to the inconsistencies of their preferred candidate, the party faithful automatically recruited brain regions that are responsible for controlling emotional reactions, such as the prefrontal cortex. While this data might suggest that voters are rational agents calmly assimilating the uncomfortable information, Westen already knew that wasn't happening, since the ratings of Kerry and Bush were entirely dependent on the subjects' party affiliations. What, then, was the prefrontal cortex doing? Westen realized that voters weren't using their reasoning faculties to analyze the facts; they were using reason to preserve their partisan certainty. And then, once the subjects had arrived at favorable interpretations of the evidence, blithely excusing the contradictions of their chosen candidate, they activated the internal reward circuits in their brains and experienced a rush of pleasurable emotion. Self-delusion, in other words, felt really good. "Essentially, it appears as if partisans twirl the cognitive kaleidoscope until they get the conclusions they want," Westen says, "and then they get massively reinforced for it, with the elimination of negative emotional states and activation of positive ones."

This flawed thought process plays a crucial role in shaping the opinions of the electorate. Partisan voters are convinced that they're rational—it's the other side that's irrational—but actually, all of us are *rationalizers*. The Princeton political scientist

Larry Bartels analyzed survey data from the 1990s to prove this point. During the first term of Bill Clinton's presidency, the budget deficit declined by more than 90 percent. However, when Republican voters were asked in 1996 what happened to the deficit under Clinton, more than 55 percent said that it had *increased*. What's interesting about this data is that so-called high-information voters—these are the Republicans who read the newspaper, watch cable news, and can identify their representatives in Congress—weren't better informed than low-information voters. (Many low-information voters struggled to name the vice president.) According to Bartels, the reason knowing more about politics doesn't erase partisan bias is that voters tend to assimilate only those facts that confirm what they already believe. If a piece of information doesn't follow Republican talking points—and Clinton's deficit reduction didn't fit the tax-and-spend liberal stereotype—then the information is conveniently ignored. "Voters think that they're thinking," Bartels says, "but what they're really doing is inventing facts or ignoring facts so that they can rationalize decisions they've already made." Once you identify with a political party, the world is edited to fit with your ideology.

At such moments, rationality actually becomes a liability, since it allows us to justify practically any belief. The prefrontal cortex is turned into an information filter, a way to block out disagreeable points of view. Let's look at an experiment done in the late 1960s by the cognitive psychologists Timothy Brock and Joe Balloun. Half of the subjects involved in the experiment were regular churchgoers, and half were committed atheists. Brock and Balloun played a tape-recorded message attacking Christianity, and, to make the experiment more interesting, they added an annoying amount of static—a crackle of white noise—to the recording. However, the listener could reduce the static by pressing a button, at which point the message suddenly became easier to understand.

The results were utterly predicable and rather depressing: the nonbelievers always tried to remove the static, while the religious subjects actually preferred the message that was harder to hear. Later experiments by Brock and Balloun that had smokers listening to a speech on the link between smoking and cancer demonstrated a similar effect. We all silence the cognitive dissonance through self-imposed ignorance.

This sort of blinkered thinking isn't a problem for only partisan voters and devout believers. In fact, research suggests that the same flaw also afflicts those people who are supposed to be most immune to such cognitive errors: political pundits. Even though pundits are trained professionals, presumably able to evaluate the evidence and base their opinions on the cold, hard facts — that's why we listen to them — they are still vulnerable to cognitive mistakes. Like partisan voters, they selectively interpret the data so that it proves them right. They'll distort their thought process until it leads to the desired conclusion.

In 1984, the University of California at Berkeley psychologist Philip Tetlock began what he thought would be a brief research project. At the time, the Cold War was flaring up again — Reagan was talking tough to the "evil empire" — and political pundits were sharply divided on the wisdom of American foreign policy. The doves thought Reagan was needlessly antagonizing the Soviets, while the hawks were convinced that the USSR needed to be aggressively contained. Tetlock was curious which group of pundits would turn out to be right, and so he began monitoring their predictions.

A few years later, after Reagan left office, Tetlock revisited the opinions of the pundits. His conclusion was sobering: *everyone was wrong*. The doves had assumed that Reagan's bellicose stance would exacerbate Cold War tensions and had predicted a breakdown in diplomacy as the USSR hardened its geopolitical stance. The reality, of course, was that the exact opposite happened. By 1985, Mikhail Gorbachev was in power. The Soviet Union began implementing a stunning series of internal reforms. The "evil empire" was undergoing *glasnost*.

But the hawks didn't do much better. Even after Gorbachev began the liberalizing process, hawks tended to disparage the changes to the Soviet system. They said the evil empire was still evil; Gorbachev was just a tool of the politburo. Hawks couldn't imagine that a sincere reformer might actually emerge from a totalitarian state.

The dismal performance of these pundits inspired Tetlock to turn his small case study into an epic experimental project. He picked 284 people who made their living "commenting or offering advice on political and economic trends" and began asking them to make predictions about future events. He had a long list of pertinent questions. Would George Bush be reelected? Would there be a peaceful end to apartheid in South Africa? Would Quebec secede from Canada? Would the dot-com bubble burst? In each case, the pundits were asked to rate the probability of several possible outcomes. Tetlock then interrogated the pundits about their thought processes so he could better understand how they'd made up their minds. By the end of the study, Tetlock had quantified 82,361 different predictions.

After Tetlock tallied the data, the predictive failures of the pundits became obvious. Although they were paid for their keen insights into world affairs, they tended to perform *worse* than random chance. Most of Tetlock's questions had three possible answers; on average, the pundits had selected the right answer less than 33 percent of the time. In other words, a dart-throwing chimp would have beaten the vast majority of professionals. Tetlock also found that the most famous pundits in his study tended to be the least accurate, consistently churning out overblown and overconfident forecasts. Eminence was a handicap.

Why were these pundits (especially the prominent ones) so bad at forecasting the future? The central error diagnosed by

Tetlock was the sin of *certainty,* which led the "experts" to mistakenly impose a top-down solution on their decision-making processes. In chapter 2, we saw examples of the true expertise that occurs when experience is internalized by the dopamine system. This results in a person who has a set of instincts that respond quickly to the situation at hand, regardless of whether that's playing backgammon or staring at a radar screen. The pundits in Tetlock's study, however, distorted the verdicts of their emotional brains, cherry-picking the feelings they wanted to follow. Instead of trusting their gut feelings, they found ways to disregard the insights that contradicted their ideologies. When pundits were convinced that they were right, they ignored any brain areas that implied they might be wrong. This suggests that one of the best ways to distinguish genuine from phony expertise is to look at how a person responds to dissonant data. Does he or she reject the data out of hand? Perform elaborate mental gymnastics to avoid admitting error? Everyone makes mistakes; the object is to learn from these mistakes.

Tetlock notes that the best pundits are willing to state their opinions in "testable form" so that they can "continually monitor their forecasting performance." He argues that this approach makes pundits not only more responsible—they are forced to account for being wrong—but also less prone to bombastic convictions, a crucial sign that a pundit isn't worth listening to. (In other words, ignore those commentators that seem too confident or self-assured. The people on television who are most certain are almost certainly going to be wrong.) As Tetlock writes, "The dominant danger [for pundits] remains hubris, the vice of closed-mindedness, of dismissing dissonant possibilities too quickly." Even though practically all of the professionals in Tetlock's study claimed that they were dispassionately analyzing the evidence —everybody wanted to be rational—many of them were actually indulging in some conveniently cultivated ignorance. Instead of encouraging the arguments inside their heads, these pundits settled on answers and then came up with reasons to justify those answers. They were, as Tetlock put it, "prisoners of their preconceptions."

## 2

It feels good to be certain. Confidence is comforting. This desire to always be right is a dangerous side effect of having so many competing brain regions inside one's head. While neural pluralism is a crucial virtue—the human mind can analyze any problem from a variety of different angles—it also makes us insecure. You never know which brain area you should obey. It's not easy to make up your mind when your mind consists of so many competing parts.

This is why being sure about something can be such a relief. The default state of the brain is indecisive disagreement; various mental parts are constantly insisting that the other parts are wrong. Certainty imposes consensus on this inner cacophony. It lets you pretend that your entire brain agrees with your behavior. You can now ignore those annoying fears and nagging suspicions, those statistical outliers and inconvenient truths. Being certain means that you aren't worried about being wrong.

The lure of certainty is built into the brain at a very basic level. This is most poignantly demonstrated by split-brain patients. (These are patients who have had the corpus callosum—the nerve tissue that connects the two hemispheres of the brain—severed. The procedure is performed only rarely, usually to treat intractable seizures.) A typical experiment goes like this: using a special instrument, different sets of pictures are flashed to each of the split-brain patient's visual fields. (Because of our neural architecture, all information about the left visual field is sent to the right hemisphere, and all information about the right visual field is sent to the left hemisphere.) For example, the right visual field might see a picture of a chicken claw and the left visual field might see a picture of a snowy driveway. The patient is then shown a variety of images and asked to pick out the one that is most closely associated with what he or she has just seen. In a tragicomic display of indecisiveness, the split-brain patient's hands point to two different objects. The right hand points to a chicken (this matches the chicken claw that the left hemisphere witnessed), while the left hand points to a shovel (the right hemisphere wants to shovel the snow). The conflicting reactions of the patient reveals the inner contradictions of each of us. The same brain has come up with two very different answers.

But something interesting happens when scientists ask a split-brain patient to explain the bizarre response: the patient manages to come up with an explanation. "Oh, that's *easy*," one patient said. "The chicken claw goes with the chicken, and you need a shovel to clean out the chicken shed." Instead of admitting that his brain was hopelessly confused, the patient wove his confusion into a plausible story. In fact, the researchers found that when patients made especially ridiculous claims, they seemed even more confident than usual. It was a classic case of overcompensation.

Of course, the self-assurance of the split-brain patient is clearly mistaken. None of the images contained a chicken shed that needed a shovel. But that deep need to repress inner contradictions is a fundamental property of the human mind. Even though the human brain is defined by its functional partitions, by the friction of all these different perspectives, we always feel compelled to assert its unity. As a result, each of us pretends that the mind is in full agreement with itself, even when it isn't. We trick ourselves into being sure.


DURING THE LAST week of September 1973, the Egyptian and Syrian armies began massing near the Israeli border. The signals picked up by the Mossad, the main Israeli intelligence agency, were ominous. Artillery had been moved into offensive positions. Roads were being paved in the middle of the desert. Thousands of Syrian reservists had been ordered to report for duty. From the hills of Jerusalem, people could see a haze of black diesel smoke on the horizon, the noxious exhaust generated by thousands of Soviet-made tanks. The smoke was getting closer.

The official explanation for the frenzy of military activity was that it was a pan-Arab training exercise. Although Anwar Sadat, the president of Egypt, had boldly declared a few months before that his country was "mobilizing in earnest for the resumption of battle" and declared that the destruction of Israel was worth the "sacrifice of one million Egyptian soldiers," the Israeli intelligence community insisted that the Egyptians weren't actually planning an attack. Major General Eli Zeira, the director of Aman, the Israeli military intelligence agency, publicly dismissed the possibility of an Egyptian invasion. "I discount the likelihood of a conventional Arab attack," Zeira said. "We have to look hard for evidence of their real intentions in the field — otherwise, with the Arabs, all you have is rhetoric. Too many Arab leaders have intentions which far exceed their capabilities." Zeira believed that the Egyptian military buildup was just a bluff, a feint intended to shore up Sadat's domestic support. He persuasively argued that the Syrian deployments were merely a response to a September skirmish between Syrian and Israeli fighter planes.

On October 3, Golda Meir, the prime minister of Israel, held a regular cabinet meeting that included the heads of Israeli intelligence. It was here that she was told about the scale of Arab preparations for war. She learned that the Syrians had concentrated their antiaircraft missiles at the border, the first time this had ever been done. In addition, several Iraqi armored divisions had moved into southern Syria. She was also informed about Egyptian military maneuvers in the Sinai that weren't part of the official "training exercise." Although everyone agreed that the news was troubling, the consensus remained the same. The Arabs were not ready for war. They wouldn't dare invade. The next cabinet meeting was scheduled for October 7, the day after Yom Kippur.

In retrospect, it's clear that Zeira and the Israeli intelligence community were spectacularly wrong. In the early afternoon of October 6, the Egyptian and Syrian armies—a force roughly equivalent to the NATO European command—launched a surprise attack on Israeli positions in the Golan Heights and Sinai Peninsula. Because Meir didn't issue a full mobilization order until the invasion was already under way, the Israeli military was unable to repel the Arab armies. Egyptian tanks streamed across the Sinai and nearly captured the strategically important Mitla Pass. Before nightfall, more than 8,000 Egyptian infantry had moved into Israeli territory. The situation in the Golan Heights was even more dire: 130 Israeli tanks were trying to hold off more than 1,300 Syrian and Iraqi tanks. By that evening, the Syrians were pressing toward the Sea of Galilee, and the Israelis were suffering heavy casualties. Reinforcements were rushed to battle. If the Golan fell, Syria could easily launch artillery at Israeli cities. Moshe Dayan, the Israeli defense minister, concluded after the third day of conflict that the chances of the Israeli nation surviving the war were "very low."

The tide shifted gradually. By October 8, the newly arrived Israeli reinforcements began to reassert control in the Golan Heights. The main Syrian force was split into two smaller contingents that were quickly isolated and destroyed. By October 10, Israeli tanks had crossed the "purple line," or the pre-war Syrian border. They would eventually progress nearly forty kilometers into the country, or close enough to shell the suburbs of Damascus.

The Sinai front was more treacherous. The initial Israeli counterattack, on October 8, was an unmitigated disaster: nearly an entire brigade of Israeli tanks was lost in a few hours. (General Shmuel Gonen, the Israeli commander of the Southern Front, was later disciplined for his "failure to fulfill his duties.") In addition, the Israeli air force had lost control of the skies; its fighter planes were being shot down at an alarming rate, as the Soviet SA-2 antiaircraft batteries proved to be much more effective than expected. ("We are like fat ducks up there," one Israeli pilot said. "And they have the shotguns.") The next several days were a tense stalemate, neither army willing to risk an attack.

The standoff ended on October 14, when Sadat ordered his generals to attack. He wanted to ease the pressure on the Syrians, who were now fighting to protect their capital. But the massive Egyptian force was repulsed—they lost nearly 250 tanks—and on October 15, the Israelis launched a successful counterattack. The Israelis struck at the seam between the two main Egyptian armies and managed to secure a bridgehead on the opposite side of the Suez Canal. This breach marked the turning point of the Sinai campaign. By October 22, an Israeli armored division was within a hundred miles of Cairo. A cease-fire went into effect a few days later.

For Israel, the end of the war was bittersweet. Although the surprise invasion had been repelled, and no territory had been lost, the tactical victory had revealed the startling fragility of the nation. It turned out that Israel's military superiority was not a guarantee of security. The small country had almost been destroyed by an intelligence failure.

AFTER THE WAR, the Israeli government appointed a special committee to investigate the *mechdal,* or "omission," that had preceded the war. Why hadn't the intelligence community anticipated the invasion? The committee had uncovered a staggering amount of evidence suggesting an imminent attack. For instance on October 4, Aman learned that, in addition to building up Egyptian and Syrian forces along the border, the Arabs had evacuated Soviet military advisers from Cairo and Damascus. The day after that, new reconnaissance photographs had revealed the movement of antiaircraft missiles to the front lines and the departure of the Soviet fleet from the port of Alexandria. At this point, it should have been clear that the Egyptian forces weren't training in the desert; they were getting ready for war.

Lieutenant Benjamin Simon-Tov, a young intelligence officer at the Southern Command, was one of the few analysts who connected the dots. On October 1, he wrote a memo urging his commander to consider the possibility of an Arab attack. That memo was ignored. On October 3, he compiled a briefing document summarizing recent aggressive Egyptian actions. He argued that the Sinai invasion would begin within a week. His superior officer refused to pass the "heretical" report up the chain of command.

Why was the intelligence community so resistant to the idea of an October attack? After the Six-Day War of 1967, the Mossad and Aman developed an influential theory of Arab strategy that they called *ha-Konseptzia* (the Concept). This theory was based largely on the intelligence of a single source in the Egyptian government. It held that Egypt and Syria wouldn't consider attacking Israel until 1975, at which point they would have an adequate number of fighter planes and pilots. (Israeli air superiority had played a key role in the decisive military victory of 1967.) The Concept also placed great faith in the Bar-Lev line, a series of defensive positions along the Suez Canal. The Mossad and Aman believed that these obstacles and reinforcements would restrain Egyptian armored divisions for at least twenty-four hours, thus allowing Israel crucial time to mobilize its reservists.

The Concept turned out to be completely wrong. The Egyptians were relying on their new surface-to-air missiles to counter the Israeli air forces; they didn't need more planes. The Bar-Lev line was easy to breach. The defensive positions were mostly made of piled desert sand, which the Egyptian military moved using pressured water cannons. Unfortunately, the Concept was deeply ingrained in the strategic thinking of the Israeli intelligence community. Until the invasion actually began, the Mossad and Aman had insisted that no invasion would take place. Instead of telling the prime minister that the situation on the ground was uncertain and ambiguous—nobody really knew if the Egyptians were bluffing or planning to attack—the leaders of the Mossad and Aman chose to project an unshakable confidence in the Concept. They were misled by their certainty, which caused them to ignore a massive amount of contradictory evidence. As the psychologist Uri Bar-Joseph noted in his study of the Israeli intelligence failure, "The need for cognitive closure prompted leading analysts, especially Zeira, to 'freeze' on the conventional wisdom that an attack was unlikely and to become impervious to information suggesting that it was imminent."

Even on the morning of October 6, just a few hours before Egyptian tanks crossed the border, Zeira was still refusing to admit that a mobilization might be necessary. A top-secret cable had just arrived from a trusted source inside an Arab government, warning that an invasion was imminent, that Syria and Egypt weren't bluffing. Meir convened a meeting with her top military officials to assess this new intelligence. She asked Zeira if he thought the Arab nations were going to attack. Zeira said no. They would not dare to attack, he told the prime minister. Of that he was certain.

THE LESSON OF the Yom Kippur War is that having access to the necessary information is not enough. Eli Zeira, after all, had more than enough military intelligence at his disposal. He saw the tanks at the border; he read the top-secret memos. His mistake was that he never forced himself to consider these inconvenient facts. Instead of listening to the young lieutenant, he turned up the static dial and clung to the Concept. The result was a bad decision.

The only way to counteract the bias for certainty is to encourage some inner dissonance. We must force ourselves to think about the information we don't want to think about, to pay attention to the data that disturbs our entrenched beliefs. When we start censoring our minds, turning off those brain areas that contradict our assumptions, we end up ignoring relevant evidence. A major general shrugs off the evacuation of Soviet military personnel and those midnight cables from trusted sources. He insists that an invasion isn't happening even when it has already begun.

But the certainty trap is not inevitable. We can take steps to prevent ourselves from shutting down our minds' arguments too soon. We can consciously correct for this innate tendency. And if those steps fail, we can create decision-making environments that help us better entertain competing hypotheses. Look, for example, at the Israeli military. After failing to anticipate the 1973 war, Israel thoroughly revamped its intelligence services. It added an entirely new branch of intelligence analysis, the Research and Political Planning Center, which operated under the auspices of the Foreign Ministry. The mission of this new center wasn't to gather more information; the Israelis realized that data collection wasn't their problem. Instead, the unit was designed to provide an assessment of the available data, one that was completely independent of both Aman and the Mossad. It was a third opinion, in case the first two opinions were wrong.

At first glance, adding another layer of bureaucracy might seem like a bad idea. Interagency rivalries can create their own set of problems. But the Israelis knew that the surprise invasion of 1973 was a direct result of their false sense of certainty. Because Aman and the Mossad were convinced that the Concept was accurate, they had ignored all contradictory evidence. Complacency and stubbornness soon set in. The commission wisely realized that the best way to avoid such certainty in the future was to foster diversity, ensuring that the military would never again be seduced by its own false assumptions.

The historian Doris Kearns Goodwin made a similar point about the benefits of intellectual diversity in *Team of Rivals,* her history of Abraham Lincoln's cabinet. She argues that it was Lincoln's ability to deal with competing viewpoints that made him such a remarkable president and leader. He intentionally filled his cabinet with rival politicians who had extremely different ideologies; antislavery crusaders, like Secretary of State William Seward, were forced to work with more conservative figures, like Attorney General Edward Bates, a man who had once been a slave owner. When making a decision, Lincoln always encouraged vigorous debate and discussion. Although several members of his cabinet initially assumed that Lincoln was weak willed, indecisive, and unsuited for the presidency, they eventually realized that his ability to tolerate dissent was an enormous asset. As Seward said, "The president is the best of us."

The same lesson can be applied to the brain: when making decisions, actively resist the urge to suppress the argument. Instead, take the time to listen to what all the different brain areas have to say. Good decisions rarely emerge from a false consensus. Alfred P. Sloan, the chairman of General Motors during its heyday, once adjourned a board meeting soon after it began. "Gentlemen," Sloan said, "I take it we are all in complete agreement on the decision here . . . Then I propose we postpone further discussion of this matter until our next meeting to give ourselves time to develop disagreement and perhaps gain some understanding of what the decision is all about."

# 8

## The Poker Hand

Michael Binger is a particle physicist at Stanford. His specialty is quantum chromodynamics, a branch of physics that studies matter in its most elemental form. Binger is also a professional poker player, and he spends most of June and July sitting at felt-lined card tables in a Las Vegas casino, competing in the World Series of Poker, the most important gambling event in the world. He is one of thousands of poker players who make the pilgrimage every year. These card sharks might not look like professional athletes — the tournament is full of overweight chain smokers — but that's because they are athletes of the *mind*. When it comes to playing poker, the only thing that separates the experts from the amateurs is the quality of their decisions.

During the days of the World Series of Poker, Binger quickly settles into a mentally exhausting routine. He begins playing cards around noon — his preferred game is Texas hold'em — and he doesn't cash in his chips until the wee hours of the morning. Then it's back to his hotel room — past the strip clubs, penny slots, and the $7.77 all-you-can-eat buffets—where Binger tries to coax himself into a few hours of fitful sleep. "You get so wired playing poker that it's not easy coming down," he says. "I tend to just lie in bed, thinking about all the hands I played and how I should have played them differently."

Binger began playing cards as a college student, when he was a math and physics major at North Carolina State University. One weekend, he decided to learn how to play blackjack. He quickly grew frustrated by the amount of luck involved—"I hated not knowing when to bet," he says—and so he taught himself how to count cards. He practiced in loud North Carolina bars so that he could learn how to focus amid the noise and revelry. Binger is blessed with a quantitative mind—"I was always the nerd who did math problems for fun," he confesses—and counting cards came naturally to him. He quickly learned how to keep a running tally in his head, giving him a crucial advantage at the table. (For the most part, Binger relied on the Hi-Lo card-counting system, which provides the player with a 1 percent edge over the house.) Before long, Binger was traveling to casinos and putting his quantifying talents to work.

"The first thing I learned from counting cards," Binger says, "is that you can use your smarts to win. Sure, there's always luck, but over the long run you'll come out ahead if you're thinking right. The second thing I learned is that you can be *too* smart. The casinos have algorithms that automatically monitor your betting, and if they detect that your bets are too accurate, they'll ask you to leave." This meant that Binger needed to occasionally make bad bets on purpose. He would intentionally lose money so that he could keep on making money.

But even with this precaution, Binger started to put numerous casinos on alert. In blackjack, it's supposed to be impossible to consistently beat the house, and yet that's exactly what Binger kept doing. Before long, he was blacklisted; casino after casino told him that he couldn't play blackjack at their tables. "Some of the casinos would ask politely," Binger says. "A manager would come and tell you to take your winnings and leave. And some casinos weren't so polite. Let's just say they made it clear you weren't welcome anymore."

After he started graduate school in theoretical physics at Stanford, Binger tried to shake his card habit. "The low point for me was getting kicked out of six Reno casinos in one day for counting cards," he says. "That's when I realized maybe I should focus on physics for a while." He was drawn to the most difficult problems in the field, studying supersymmetry and the Higgs boson particle. (The elusive Higgs is often referred to as the "God particle," since finding it would help explain the origin of the universe.) "There's no doubt that the analytical skills I learned in cards also helped me with science," Binger says. "It's all about focusing on the important variables, thinking clearly, not getting distracted. If you lose your train of thought when you're counting cards, you're screwed. Physics is a little more forgiving—you can write stuff down—but it still requires a very disciplined thought process."

After a few years of working diligently toward his PhD, Binger began to miss his beloved card games. The relapse was gradual. He started playing a few rounds of small-stakes poker with his friends, just a casual game or two after a long day spent contemplating physics equations. But it didn't take long before Binger's friends refused to play with him; he kept taking all of their money. And so Binger began playing poker tournaments, driving out on the weekends to the card rooms near the San Francisco airport. After a few months, Binger was making more money on the part-time poker circuit than he made as a postdoc. He used his winnings to pay off his student loans and start a modest bankroll. "I realized that I'd never be able to really focus on physics until I gave poker a shot," he says. "I needed to know if I could make it." That's when Binger decided to try his luck as a professional gambler.

THE WORLD SERIES of Poker (WSOP) is held at the Rio Hotel, a Brazilian-themed casino that's across the highway from the Strip. For the most part, the Latin motif is confined to the silly costumes of the staff, the syrupy cocktails, and the ugly carpets, which are a swirl of Carnival colors. The hotel itself is a generic tower of reflective purple and red glass. During the WSOP, the lobby of the Rio accumulates the litter of the tournament: cigarette butts, empty water bottles, registration papers, fast-food wrappers. Anxious players collect in the corners, sharing stories of bad beats and lucky breaks. Even the hotel gift shop is stocked for the event, carrying a wide selection of poker primers right next to the nudie magazines.

Most of the tournament takes place in the Amazon Room, a cavernous warehouse-like space with more than two hundred card tables. Security cameras dangle from the ceiling like ominous disco balls. Compared with the rest of Vegas, the atmosphere inside the huge room is starkly sober and serious. (Nobody would dare litter in here.) Even when it's filled with poker players, the enormous area has moments of startling calm, when all you can hear are the shuffling of cards and the perpetual hum of the air-conditioning system. Outside, it's 114 degrees.

Binger is tall and lean, with a face made of angles. His hair is boyishly blond, and it's usually styled with copious amounts of gel so that it sticks straight up. At every poker tournament, he wears the same outfit: a backward baseball cap, opaque Oakley sunglasses, and a brightly colored button-down shirt. Such consistency is typical of poker players, who are creatures of habit and rigid believers in routine. (A common quip on the pro circuit is "It's unlucky to be superstitious.") Some professionals wear the same sweatshirts day after day, until the reek of their nervousness precedes them. Others develop bizarre eating rituals, like Jamie Gold, who orders scrambled eggs for breakfast even though he's allergic to eggs.

Binger actually eats his eggs. His breakfast routine consists of one egg over easy, sandwiched between a lightly toasted English muffin. After eating that, he drinks a small glass of orange juice and then a strong cup of tea. He digests for "approximately ten to twelve minutes," and next drives to the gym, where he performs an extremely regimented workout. "All these habits probably sound a little crazy," Binger says, "but when you're playing in a tournament it's crucial to not distract yourself with thinking about what to order for breakfast or how many laps to swim. The routine keeps it simple, so all I'm thinking about is poker, poker, poker."

At the 2006 WSOP, Binger was one of 8,773 players who each paid $10,000 to enter the main event, a no-limit Texas hold'em competition stretching over thirteen days. Since 1991, when the prize money for the WSOP first exceeded a million dollars, the poker tournament has been more lucrative for its winners than Wimbledon, the PGA championship, and the Kentucky Derby. Since 2000, it has become the most valuable sporting event in the world, at least for the winners. (More than 90 percent of entrants won't "make the money," which means that they'll lose their entire entry fee.) In 2006, the top prize for the main event was expected to exceed twelve million dollars. To make an equivalent amount of money playing tennis, you'd have to win Wimbledon ten times.

The rules of Texas hold'em are simple. Nine players gather around a card table, each of them determined to assemble the best possible poker hand. The game begins when each player is dealt two cards, face-down. The two players to the left of the dealer are then forced to make blind bets, wagering their money before they even look at their cards. (These bets ensure that there's some money at stake in every hand.) The remaining players have three options: they can match the bet, raise it, or fold. If a player has strong hole cards—a pair of aces being the best possible duo—he or she will make an aggressive bet. (Unless, of course, the player wants to act weak, but that's another story.) A bad hand is a good reason to fold.

After the first round of betting is over, three community cards are dealt, face-up, in the center of the table. These cards are called the flop. There is now another round of betting, as players adjust their wagers in light of this new information. Then two more community cards are dealt, one at a time, with another round of betting after each. (The fourth card is called the turn, and the fifth card is called the river.) Each player then assembles the most valuable hand possible by combining the two hole cards with any three of the five community cards shared by the entire table. So let's say you're dealt the ace and the ten of hearts. The best possible set of community cards would contain the jack, queen, and king of hearts, since that would give you a royal flush, the perfect poker hand. (A royal flush is dealt approximately once every 648,739 poker hands.) If you got the jack, queen, and king of different suits, then you'd have a straight. (Odds: 253 to 1.) You'd also be thrilled with three heart cards of any value, since that would give you a flush. (Odds: 507 to 1.) A more likely scenario is that you end up with a single pair (odds are 1.37 to 1), or you might get absolutely nothing, in which case the highest-ranking card, the ace, is your entire hand.

At its core, poker is a profoundly statistical game. Each hand is ranked according to its rarity, so having two pairs is more valuable than having one, and a straight flush is more valuable than a straight or a flush. A poker player who can parse his hole cards into possible probabilities—who knows, for example, that being dealt a pair of fours means that there's a 4 percent chance of getting dealt another four on the flop—has a distinct advantage over his competitors. He can make bets informed by the laws of statistics, so that his wagers reflect the likelihood of winning the hand.

But the game isn't just about the cards. The act of betting is what makes poker so infinitely complicated. It's what turns Texas hold'em into a black art, a mixture of stagecraft and game theory. Consider the act of raising the bet. Such a move can have a straightforward meaning: a player is demonstrating confidence in his hole cards. Or it can signal a bluff, as a player tries to steal the pot by intimidating all the other players into folding. How does one distinguish between these different intentions? That's where the skill comes in. Professional poker players are constantly trying to read their opponents, searching for the minor tells of deceit. Does this bet fit into a behavioral pattern? Has the player been consistently "tight" or aggressive all night? Why is his left eye twitching? Is that a symptom of nervousness? (Those who are easy to read are known as ABC players.) Of course, the best poker players are also the best liars, able to keep their opponents off balance with sincere bluffs and unpredictable bets. They know that the most important thing in poker is not what cards they actually have, but what cards people *think* they have. A lie told well is just as good as the truth.

IN THE BEGINNING of the tournament, Binger played patient poker, using his extraordinary math skills—a talent he honed in grad school—to methodically figure out which hands he should enter. Nine times out of ten he immediately folded, and he risked his money only when he had hole cards with decent statistical odds, such as a high pair or an ace-king combo. "The opening rounds of every tournament are always full of players who probably shouldn't be there," Binger says. "These are the rich guys who think they are much better than they actually are. At this stage of the game, the most important thing you can do is not make a big mistake. You don't want to take unnecessary risks. You just want to stay alive. This is when I'm making sure that I'm always doing the math."

Look, for example, at one of Binger's early hands at the WSOP. He was dealt an immaculate pair of aces, a hand so good it has its own name (it's called American Airlines). Naturally, Binger decided to raise. Although it was a modest raise—Binger didn't want to scare anybody off—everyone at the table decided to fold, except for a well-groomed older man wearing a canary yellow polo shirt with big sweat stains in the armpits. He pushed his short stack of chips to the center of the table. "I'm all in," said the man in yellow. Binger assumed that the man had either a high pair (like two kings) or two high cards of the same suit (like the king and queen of spades). Binger paused for a moment and contemplated his odds. If he correctly read the other player's hand—and that was a big *if*—then he had somewhere between an 82 and 87 percent of winning. Binger decided to match the bet. The man nervously turned over his cards: the ace and jack of diamonds. The flop was dealt but it was a meaningless collection of number cards. The turn and the river were more of the same. Binger's pair of aces prevailed. The man in the yellow shirt winced and walked off without a word.

As the days pass, the weak players are ruthlessly culled from the tournament. It's like natural selection on fast-forward. The tournament doesn't end for the night until more than half of the players have been eliminated, so it's not unusual for the nights to last until two or three in the morning. ("Learning how to become nocturnal is part of the challenge," Binger says.) By the fourth day, even the skilled survivors are beginning to look a little worn out from the struggle. Their faces are masks of fatigue and stubble, and their eyes have the faraway look of an adrenaline hangover. The smell of stale cigarette smoke seems to be a popular deodorant.

Binger is gradually getting more aggressive at the poker table.

It's as if his betting instincts have a dial, and he's slowly turning up the volume. He's still folding the vast majority of his hands, but when he does decide to make a bet, he doesn't equivocate. In these situations, his table manners follow a well-rehearsed script. Binger takes a second glance at his hole cards and flexes his jaw muscles. He then adjusts his reflective sunglasses, pressing them tight against his eyes, looks at his cards again, and pushes an intimidating pile of chips to the center of the table. His face radiates self-assurance. He's done the calculations and knows the odds. Most of the time, the other players respond by folding.

This fiercely disciplined strategy pays off. By the end of the fifth day of the tournament, Binger is in fourth place, with $4,920,000 in chips. Fourteen hours later, he's got $5,275,000. After seven exhausting days, he's amassed a pile of nearly $6,000,000 in chips. And then, on the eighth day, Binger makes the final table. When play begins, the Hollywood producer Jamie Gold has a commanding chip lead over the other players. Gold has been playing smart poker, but he's also been enjoying a staggering run of good luck. As one poker professional later told me, "[Gold] has an amazing ability to pull cards out of his ass. And somehow he always pulls the exact right card."

After a few hours, Gold begins to eliminate some of the remaining players. His big chip lead means that he can turn each hand into a potential trap. Gold can also bluff with abandon, since calling his bluff means the other player has to go all in. Binger is playing conservatively — "I just wasn't getting the right cards," he said later — and so he waits, and watches. The big antes are gradually bleeding away his chips, but he's getting a better sense of his competition. "After a while, you just get these feelings about people," he says. "You'll watch them make a certain bet and then scratch their nose or whatever and all of a sudden you'll realize that they've got nothing, that you can take the hand." There are no certainties in poker, which means that anything that can narrow the uncertainty is extremely valuable, even if it's just a subtle hunch. These psychological interpretations aren't quantifiable—you can't summarize a person in a probability—but they still inform Binger's betting decisions.

When there are only five players left, Binger begins to make his move. "It started when I got dealt a pair of kings," he says. "I decided right then and there to make a rather aggressive bet." A few hours before, Binger had bluffed one of the five, a player named Paul Wasicka, out of a big pot. Although Binger had been dealt a poor hand, his aggressive bet convinced everyone else to fold. Binger could tell that Wasicka was still seething. "I knew Paul thought I was just trying to bluff him again," Binger remembers. "He thought that I only had a small pair. But I had pocket kings."

Binger wanted to draw Wasicka deeper into the hand. In tactical moments like this, poker transcends its probabilities. The game morphs into a deeply human drama, a competition of decision-making. Binger needed to make a bet that would convince Wasicka he was trying to steal another pot, that he was once again making an aggressive bet with nothing but a low pair. "I decided to go all in," Binger says. "By overplaying my hand—by pretending to act strong—I was actually acting weak, at least in his eyes. I then tried to exude weakness, but without making it obvious, because then he would know that I was only *pretending* to bluff, which is a sure sign that I've actually got a good hand." Binger's best friend and brother were both watching the tournament on closed-circuit television. The best friend was convinced that Binger was bluffing and that he was about to get knocked out of the tournament. The signs of repressed anxiety were unmistakable: Binger's fingers were manically tapping on the table, and his teeth were digging into his lower lip. "Only my brother knew better," Binger says. "I guess he knew how to read my face. He said I looked too weak, so I must be strong."

Wasicka took the bait. He was so certain that Binger was bluffing that he ended up betting millions of chips on a weak hand. Binger won the pot and doubled his chips. "That bet had nothing to do with math," Binger says. "I'd gotten high pairs before, and not done much with them . . . But at that moment, as soon as I saw my cards, I knew what I needed to do. To be honest, I don't know why I went all in on that hand. If I'd really thought about it, I might not have done it. The bet was damn risky. But it just felt like the right thing to do. You can do all the probabilistic analysis in the world, but in the end it all comes down to something you can't quite explain."

# 1

Professional poker players are a fatalistic bunch. They live in a deterministic world shaped by mysterious forces. Everything is possible, and yet only one thing ever happens. You might get the card you need on the river, but you might not. There's a possibility you'll make the straight, but you probably won't. Poker is a game of subtle skill and exquisite odds, but it's also a crapshoot.

This undercurrent of chance is the defining feature of the game. It's what makes the psychological aspects of poker—the subtle reads, the convincing bluffs, the inexplicable intuitions—so essential. Chess, by contrast, is a game of pure information. There are no secrets or shuffled decks or hidden cards; the moving parts of the game are all perfectly visible, right there on the chessboard. As a result, computers can consistently beat grand masters; they can use their virtually unlimited processing powers to find the perfect move. But poker isn't so amenable to microchips and mathematics. Great poker players aren't just gambling statisticians. They need to bring something else to the table, to possess that inexplicable talent for knowing when to risk everything on a pair of kings. "Poker is a science, but it's also an art," Binger says. "To be good, you have to master both sides of the game."

What Binger is alluding to is the fact that there are always two ways to look at a poker hand. The first approach is mathematical. It treats every hand like a math problem and assumes that winning the game is simply a matter of plugging the probabilities into a sophisticated equation. According to this strategy, poker players should act like rational agents, looking for bets that minimize risk and maximize gain. This is what Binger did during the opening rounds of the WSOP, when he was only betting on high-percentage hands. Making money was just a matter of getting the odds right.

But Binger knows that poker isn't merely a set of math problems. When he talks about the art of the game, he's alluding to everything that can't be translated into numbers. The laws of statistics couldn't have told Binger how to lead Wasicka into his trap, or whether he should bluff with a middling pair. Even the most carefully calculated odds can't eliminate the unpredictability in a shuffled deck of cards. This is why the best poker players don't pretend that poker can be solved. They know the game is ultimately a *mystery.*

The difference between math problems and mysteries is important. In order to solve a math problem, all you need is more rational thought. Some poker hands, of course, can be played by relying on the math: if you're dealt a pair of aces, or get a straight on the flop, then you're going to make an aggressive bet. The odds are in your favor, and a little statistics will lead you to make the correct decision. But this rational approach can't be applied to the vast majority of poker hands, which are utter mysteries. In these situations, more statistical analysis won't help the player make a decision. In fact, thinking too much is part of the problem, since all that extra thought just gets in the way. "Sometimes, I have to tell myself to not focus on the math," Binger says. "The danger with the math is that it can make you think you know more than you do. Instead of thinking about what the other player is doing, you end up obsessing over the percentages." The first part of solving a mystery is realizing that there is no easy solution. Nobody knows what card is coming next.

This is where feelings come into play. When there is no obvious answer, a poker player is forced to make a decision using the emotional brain. And so that vague intuition about his hand, that inexplicable hunch about his opponent, ends up becoming a decisive factor. This decision won't be perfect—there's too much uncertainty for that—but it's the best option. Mysteries require more than mere rationality. "I know that my mind assimilates many more variables than I'm actually aware of," Binger says. "Especially when it comes to reading other players, I'll often make strong and accurate reads without knowing what signals I'm picking up on. And as I've gained experience, I've felt my poker instincts just get better and better, to the point where I almost never doubt them. If I get a strong feeling, then that's what I go with."

Remember Damasio's card-playing experiment? In that gambling game, players had to turn over about eighty cards before they could consciously explain which card deck was the best option. Their conclusions were rational, but they were also rather slow. It takes a while to do the math. But when Damasio measured people's emotions, he discovered that their feelings were able to identify the good decks after only ten cards. Whenever people reached for the risky decks, they experienced a surge of nervousness, even though they couldn't say why they felt so nervous. The subjects who trusted their emotional brains—who listened to their sweaty palms—made the most money.

The different strategies used by poker players illuminate the benefits of having a mind capable of rational analysis *and* irrational emotion. Sometimes it helps to look at cards from the cold perspective of statistics, to bet on hands only when the odds are on your side. But the best poker players also know when *not* to rely on the math. People aren't particles. To play the game is to accept the limits of statistics, to know that numbers don't know everything. Binger realizes that in certain situations, it's important to listen to his feelings, even if he doesn't always know what they're responding to. "As a physicist, it can be hard admitting that you just can't reason your way to the winning hand," Binger says. "But that's the reality of poker. You can't construct a perfect model of it. It's based on a seemingly infinite amount of information. In that sense, poker is a lot like real life."

## 2

Ap Dijksterhuis, a psychologist at the University of Amsterdam, had a scientific breakthrough while shopping for a car. Like most consumers, Dijksterhuis was slightly overwhelmed by the variety of makes and models. There were just so many alternatives to consider. Before he could find the right car, Dijksterhuis needed to take a dizzying number of variables into account, from fuel economy to trunk space. And then, once he made up his mind, Dijksterhuis had to figure out which options he wanted. A moon roof? A diesel engine? Six speakers? Side air bags? The list of possibilities seemed endless.

That's when Dijksterhuis realized that buying a car exceeded the limits of his conscious brain. He could no longer remember whether the Toyota or the Opel had a bigger engine, or if it was the Nissan or the Renault that offered the attractive lease. All the different variables were blurred together; his prefrontal cortex was confused.

But if Dijksterhuis couldn't keep track of the different cars, then how could he ever make a decision? Was he destined to pick the wrong car? What was the best way to make a difficult choice? To answer these questions, Dijksterhuis decided to conduct a practical experiment; it was later published in *Science*. He got several Dutch car shoppers and gave them each descriptions of four different used cars. Each of the cars was rated in four different categories, for a total of sixteen pieces of information. Car number 1, for example, was described as getting good mileage but having a shoddy transmission and a poor sound system. Car number 2 handled poorly but had lots of legroom. Dijksterhuis designed the experiment so that one car was objectively ideal, with "predominantly positive aspects." After showing a subject these car ratings, Dijksterhuis would give him a few minutes to contemplate the decision. In this "easy" situation, more than 50 percent of the subjects ended up choosing the best car.

Dijksterhuis then showed a different selection of people the same car ratings. This time, however, he didn't let each of them consciously think about the decision. After he gave the automotive facts, he distracted the subject with some simple word games for a few minutes, then interrupted the fun and suddenly asked the person to choose a car. Dijksterhuis had designed the experiment so that the person would be forced to make a decision using the unconscious brain, by relying on his or her emotions. (Conscious attention had been focused on solving the word puzzle.) The result was that these subjects made significantly worse choices than those who were allowed to consciously think about the cars.

So far, so obvious. A little rational analysis could have prevented the "unconscious choosers" from buying a bad car. Such data confirms the conventional wisdom: reason is always better. We should think before we decide.

But Dijksterhuis was just getting warmed up. He repeated the experiment, only this time he rated each car in *twelve* different categories. (These "hard" conditions more closely approximate the confusing reality of car shopping, in which consumers are overwhelmed with facts and figures.) In addition to getting information about the quality of the transmission and the engine's gas mileage, people were told about the number of cup holders, the size of the trunk, and so on. Their brains had to deal with forty-eight separate pieces of information.

Did conscious deliberation still lead to the best decision? Dijksterhuis found that people who were given time to think in a rational manner—those who could carefully contemplate each alternative—now chose the ideal car less than 25 percent of the time. In other words, they performed *worse* than random chance. However, subjects who were distracted for a few minutes—those who were forced to choose with their emotions—found the best car nearly 60 percent of the time. They were able to sift through the clutter of automotive facts and find the ideal alternative. The best car was associated with the most positive feelings. These irrational choosers were the best decision-makers by far.

But perhaps this data is an artifact of the lab, an effect of making people choose cars under artificial conditions. So Dijksterhuis ventured out into the real world. He began by surveying shoppers in a variety of different stores, asking them what information they considered when making their decisions. Based on these responses, he assigned a "complexity score" to a list of consumer products. Dijksterhuis found that some products, such as cheap kitchen tools (can openers, vegetable peelers, oven mitts, and so on) and home accessories (light bulbs, toilet paper, umbrellas, and so on), were relatively easy for shoppers to select. People didn't weigh many variables when making up their minds, because there weren't that many variables to consider. Since most stores carried only a few different brands of vegetable peelers and toilet paper, shoppers were able to quickly focus on the most important factors, like price. Making these simple consumer choices was the equivalent of choosing a car after learning only four attributes.

And, sure enough, when Dijksterhuis studied people shopping for modest cooking accessories, he discovered that spending more time thinking about their decisions led to more satisfaction later on. In general, people did best when they carefully compared all of their options and reasoned their way to the best vegetable peelers. They tended to regret their impulse purchases, since they'd end up with kitchen tools they didn't want or like. When buying easy consumer products, it's a good idea to take a few moments and reflect on the purchase.

Dijksterhuis then studied a more complicated shopping experience. His survey found that choosing furniture is one of the hardest consumer decisions, since it involves so many different variables. Consider a leather couch. First, you need to figure out if you like the way it looks and feels. (As Timothy Wilson demonstrated with strawberry jam, simply deciphering one's own preferences can be a very difficult cognitive task.) Then, you need to think about whether the couch will work at home. Will it clash with the coffee table? Match the drapes? Will the cat scratch the leather? Before you can make a good decision about the couch, this long list of questions needs to be answered. The problem is that the prefrontal cortex can't handle this much information by itself. As a result, it tends to fix on just one variable that may or may not be relevant, such as the color of the leather. The rational brain is forced to oversimplify the situation. Look, for instance, at the doctors who relied on MRIs to diagnose the causes of back pain; because the MRI provided them with so much anatomical data, they ended up focusing on spinal disc abnormalities, even though these abnormalities probably weren't the cause of the pain. This resulted in a lot of unnecessary surgery.

After shadowing shoppers at IKEA, the furniture warehouse, Dijksterhuis found that the longer people spent analyzing their options, the *less* satisfied they were with their decisions. Their rational faculties had been overwhelmed by the furniture store, and they ended up choosing the wrong leather couch. (IKEA offers more than thirty different kinds of sofas.) In other words, furniture shoppers did best when they didn't think at all and just listened to their emotional brains.

Remember the experiment involving the fine-art posters and the funny cat posters? In that study, led by Timothy Wilson, subjects were less satisfied with their choices when they consciously thought about what to choose; analyzing their own preferences caused them to misinterpret those preferences. Wilson concluded that for selecting things like posters or strawberry jam, people are better off listening to their initial instincts. One of Dijkster-huis's most recent experiments involved replicating Wilson's study, but with a twist: he wanted to see if letting people engage in *un*conscious decision-making—they looked at posters and then were distracted by a series of anagrams for seven min-utes—could lead them to make even better decisions.

The answer, it turns out, is a resounding yes. Consciously contemplating the posters once again led to the worst decisions —these people were the least satisfied with their choices when they were interviewed three weeks later. But the most satisfied subjects were those who let the poster options marinate in their unconscious brains for several minutes and then chose on the basis of which poster was associated with the most positive emo-tions. Dijksterhuis speculates that art posters benefit from such subterranean thought processes because they are complex choices requiring people to interpret their own subjective desires. It's not easy to figure out if you prefer van Gogh to Rothko, or if you'd rather look at an Impressionist landscape than an abstract ex-pressionist canvas. "Imagine being at an art auction in Paris," Dijksterhuis says. "There's a Monet for sale for a hundred mil-lion, and a van Gogh for a hundred and twenty-five million. How should we make this choice? The best strategy may be the following: First, take a good look at both of the paintings. Then leave the auction and distract yourself for a while (which is easy to do in Paris), and only then decide."

These simple experiments shed light on a very common prob-lem in everyday life. We often make decisions on issues that are exceedingly complicated. In these situations, it's probably a mis-take to consciously reflect on all the options, as this inundates the prefrontal cortex with too much data. "The moral of this research is clear," Dijksterhuis says. "Use your conscious mind to acquire all the information you need for making a decision. But don't try to analyze the information with your conscious mind. Instead, go on holiday while your unconscious mind digests it. Whatever your intuition then tells you is almost certainly going to be the best choice." Dijksterhuis argues that this psychological principle has far-reaching consequences and can also be applied to decisions that don't involve shopping. Anyone who is constantly making difficult decisions, from corporate executives to poker players, can benefit from a more emotional thought process. As long as someone has sufficient experience in that domain—he's taken the time to train his dopamine neurons—then he shouldn't spend too much time consciously contemplating the alternatives. The hardest calls are the ones that require the most feeling.

At first glance, this idea might be a little difficult to accept. We naturally assume that such choices require the analytical rigor of the rational brain. When trying to decipher a complicated situation, we believe that we need to consciously reflect on our options, to carefully think through the different car models or compare all of the possible couches at IKEA. Simple situations, on the other hand, are generally deemed suitable for emotions. You might trust your gut to choose a main course for dinner, but you wouldn't dream of letting it select your next car. That's why the average American spends *thirty-five* hours comparing automotive models before he or she makes a decision about which car to purchase.

But the conventional wisdom about decision-making has got it exactly backward. It is the easy problems—the mundane math problems of daily life—that are best suited to the conscious brain. These simple decisions won't overwhelm the prefrontal cortex. In fact, they are so simple that they tend to trip up the emotions, which don't know how to compare prices or compute the odds of a poker hand. (When people rely on their feelings in such situations, they make avoidable mistakes, like those due to loss aversion and arithmetical errors.*) Complex problems, on the other hand, require the processing powers of the emotional brain, the supercomputer of the mind. This doesn't mean you can just blink and know what to do—even the unconscious takes a little time to process information—but it does suggest that there's a better way to make difficult decisions. When choosing a couch, or holding a mysterious set of cards, always listen to your feelings. They know more than you do.

## 3

Michael Binger started winning poker tournaments once he realized that the game was more than a math problem. Although he's a physicist, trained to spot the quantitative pattern in the most stochastic of systems, Binger eventually discovered that he couldn't just crunch the numbers and expect to win the hand. He also needed to know when the numbers weren't enough. "I've been able to figure out the odds of poker hands for a while now, and yet, until recently, I never did very well in the World Series," Binger says. "I guess what you get better at is everything else, all the stuff that can't be quantified."

This epiphany allowed Binger to see the card game as it was, not as he wanted it to be. He no longer pretended that there was some universal solution to the problem of poker. The game was too complicated and unpredictable to summarize with statistics. Binger came to understand that different situations required dif-

---

*Your automatic brain is terrible at crunching numbers, which is why Binger always reflects on his poker probabilities. Consider this question: "A bat and ball cost $1.10 in total. The bat costs a dollar more than the ball. How much does the ball cost?" Your first instinct is probably ten cents, but that's the wrong answer, since it would add up to $1.20. The correct answer is five cents, but arriving at this answer requires a little conscious deliberation.

ferent modes of thought. Sometimes he had to play the odds. And sometimes he had to trust his gut.

This insight doesn't apply to poker alone. Look, for instance, at the financial markets. Wall Street is often compared to games of chance—like Vegas, it's a place where luck can be as important as logic—and when it comes to decision-making, the parallels can be illuminating. Both poker and investing are inherently unpredictable enterprises, requiring people to act with incomplete information. Nobody knows how the market will respond to the latest economic data or what card will appear on the river. Nobody knows if the Federal Reserve is going to lower interest rates next quarter or if the player with the big pile of chips is bluffing. In such situations, the only way for anyone to succeed over the long term is to use both brain systems in their proper contexts. We need to think *and* feel.

A few years ago, Andrew Lo, a business professor at MIT, wired ten currency speculators and stock traders at a brokerage firm with sensors that monitored their heart rates, blood pressure, body temperature, and skin conductivity. These bodily signs correlate with emotions: intense feelings make for fast pulses. By the end of the day, the traders had made more than a thousand financial decisions, wagering over forty million dollars. If these professional investors were perfectly rational agents, as economic theory assumes, then they should have had perfectly calm bodies. When Lo looked at the data, however, he found that the decisions of the traders were the stuff of sweaty palms and spiking blood pressure. Most financial transactions were accompanied by surges of feeling.

This wasn't necessarily a bad thing. The vast majority of emotional decisions turned out to be profitable. Just because traders had nervous hands, or frightened amygdalas, didn't mean they were acting "irrationally." Rather, Lo discovered that the traders made the worst decisions when their emotions were either silent or overwhelming. In order to make the right investment decisions, the mind needs emotional input, but those emotions need to exist in a dialogue with rational analysis. Investors who got too worked up or who tried to rely on logic alone tended to make dramatic mistakes. "One of the implications of our experiments," Lo says, "is that strong emotional reactions to financial gains or losses can actually be counterproductive. On the other hand, too little emotional reaction can also be dangerous. There's an ideal range of emotional responses that professional securities traders seem to exhibit, and that's an insight that we think individual investors can benefit from." The best investors, like the best poker players, are able to find that crucial mental balance. They constantly use one brain system to improve the performance of the other.

Just look at Binger. On the one hand, he's always using his prefrontal cortex to interrogate his emotions, to consciously question his unconscious brain. This doesn't mean he's ignoring his feelings — he's not making the strawberry-jam mistake — but it does mean that he's making sure to avoid any obvious emotional errors, what poker players refer to as tilt. "The way I look at it," Binger says, "is that it can't hurt to think for a few seconds about what I'm feeling. Most of the time, I'll still go with my instincts, but occasionally I'll catch myself doing something dumb."

Consider this poker hand from the first day of the tournament. Binger was trying to play it safe, but he ended up losing a big pile of chips when someone beat his pair of jacks on the river. Fortunately, Binger was self-aware enough to realize that such losses can trigger a dangerous set of feelings as the effects of loss aversion settle in. "You want your chips back," Binger says, "and that's when you find yourself taking risks you probably shouldn't take." At moments like this, Binger's prefrontal cortex reasserts control of his gambling decisions, preventing him from making an impulsive mistake: "I'll remind myself to play tight, to focus on the odds." You don't go all in if you've only got one out.

Situations like this demonstrate the importance of the prefrontal cortex. The rational parts of the brain are uniquely able to monitor feelings, using the reins of cognition to keep the horses from running wild. Ironically, it's those moments when emotions seem most persuasive — when the brain is completely convinced that it's time to go all in — that you should take a little extra time to reflect on the emotional decision. Make yourself consider alternative possibilities and scenarios. This is why the Israeli intelligence services added yet another analytical branch after the Yom Kippur War. "If the game seems simple or obvious, then you've made a mistake," Binger says. "The game is never simple. You've always got to wonder: what am I missing?"

Binger's ability to alternate between emotions and rationality has one important effect: it forces him to always think about *how* he's thinking. Because Binger has an array of cognitive strategies to choose from, he is constantly reflecting on which strategy he should use at any particular moment. This sort of mental flexibility is an essential feature of good decision-making. Look at the Philip Tetlock study of political pundits that we talked about in the last chapter. Although the study is best known for its demonstration of expert failure — the vast majority of pundits failed to predict better than random chance — Tetlock also found that a few performed far above average.

Tetlock explained the difference between successful and unsuccessful pundits with an allusion to an ancient metaphor made famous by historian Isaiah Berlin in his essay "The Hedgehog and the Fox." (Berlin's title is a reference to the ancient Greek expression "The fox knows many things, but the hedgehog knows one big thing.") In that essay, Berlin distinguished between two types of thinkers, hedgehogs and foxes, and Tetlock used those same categories to describe the pundits' methods of decision-making. (Tetlock did not find any significant correlation between political ideology and thinking style.) A hedgehog is a small mammal covered with spines; when attacked, it rolls itself into a ball so that its spines point outward. This is the hedge-hog's only defense. A fox, on the other hand, doesn't rely on a single strategy when threatened. Instead, it adjusts its strategy to fit the particulars of the situation. Foxes are also cunning hunt-ers. In fact, they are one of the hedgehog's few predators.

According to Tetlock, the problem with a pundit who thinks like a hedgehog is that he is prone to bouts of certainty—the big idea is irrefutable—and this certainty causes him to misinterpret the evidence. If the amygdala contradicts one of his conclusions—it's worrying about some bit of evidence that doesn't support the pundit's accepted worldview—then the amygdala is turned off. A diversity of brain regions isn't brought to bear on the problem. Useful information is deliberately ignored. The inner argument is badly argued.

A successful pundit, on the other hand, thinks like a fox. While the hedgehog reassures himself with certainty, the fox re-lies on the solvent of doubt. He is skeptical of grand strategies and unifying theories. The fox accepts ambiguity and takes an ad hoc approach when coming up with explanations. The fox gathers data from a wide variety of sources and listens to a diver-sity of brain areas. The upshot is that the fox makes better pre-dictions and decisions.

But being open-minded isn't enough. Tetlock found that the most important difference between fox thinking and hedgehog thinking is that the fox thinker is more likely to *study his own decision-making process.* In other words, he thinks about how he thinks, just like Binger.* According to Tetlock, such introspec-tion is the best predictor of good judgment. Because foxes pay attention to their inner disagreements, they are less vulnerable to

---

*Patients who have undergone cognitive-behavioral therapy (CBT), a form of talk therapy designed to reveal the innate biases and distortions of the human brain, have also been shown to be less vulnerable to these same biases. Scientists speculate that these patients have learned to recognize those maladaptive thoughts and emo-tions that automatically occur in their responses to certain situations. Because they reflect on their thought processes, they learn to think better.

the seductions of certainty. The fox doesn't tune out his insula or ventral striatum or nucleus accumbens just because it contradicts his preconceptions. "We need to cultivate the art of self-overhearing," Tetlock says, "to learn how to eavesdrop on the mental conversations we have with ourselves."

That's also the lesson of Michael Binger's success. Although Jamie Gold went on to win the 2006 WSOP, Binger's third-place finish earned him a consolation prize of $4,123,310. The next year, in the 2007 WSOP, Binger tied the all-time record for most cashes in a single tournament. (To *cash* means to win money.) He began 2008 by winning one of the main no-limit Texas hold'em events at the LA Poker Classic, earning another six-figure payday. He is now regarded as one of the best players on the professional poker circuit. "What I love about poker," Binger says, "is that when you win, it's always for the same reason. You might lose because you got unlucky, but you never win because of luck. The only way to win is to make better decisions than everyone else at the table."

## 4

We can now start to sketch out a taxonomy of decision-making, applying the knowledge of the brain to the real world. We've seen how the different brain systems — the Platonic driver and his emotional horses — should be used in different situations. While reason and feeling are both essential tools, each is best suited for specific tasks. When you try to analyze a strawberry jam or feel your way to a vegetable peeler, you are misusing your machine. When you're certain that you're right, you stop listening to those brain areas that say you might be wrong.

The science of decision-making remains a young science. Researchers are just beginning to understand how the brain makes up its mind. The cortex remains a mostly mysterious place, an extraordinary yet imperfect computer. Future experiments will reveal new aspects of human hardware and software. We'll learn about additional programming bugs and cognitive talents. The current theories will undoubtedly get complicated. And yet, even at the dawn of this new science, it's possible to come up with a few general guidelines that can help us all make better decisions.

SIMPLE PROBLEMS REQUIRE REASON. There isn't a clear line separating easy questions from hard ones, or math problems from mysteries. Some scientists, such as Ap Dijksterhuis, believe that any problem with more than four distinct variables overwhelms the rational brain. Others believe that a person can consciously process somewhere between five and nine pieces of information at any given moment. With practice and experience, this range can be slightly expanded. But in general, the prefrontal cortex is a sharply constrained piece of machinery. If the emotional brain is a fancy laptop, stuffed full of microprocessors operating in parallel, the rational brain is an old-fashioned calculator.

That said, a calculator can still be a very useful tool. One of the drawbacks of emotions is that they contain a few obsolete instincts that are no longer suited for modern life. This is why we are all so vulnerable to loss aversion, slot machines, and credit cards. The only way to defend against such innate flaws is to exercise reason, to fact-check feelings with a little arithmetic. Remember Frank, the unlucky contestant on *Deal or No Deal*? If he'd taken the time to rationally evaluate the offer, to plug the proposal into a calculator, he would have ended up with €10,000. Instead, he walked away with €10.

Of course, it's not always obvious which decisions are simple. Picking a strawberry jam or breakfast cereal might seem like an easy task, but it's actually surprisingly complicated, especially when a typical supermarket stocks more than two hundred different varieties of each. So how can anyone reliably identify the simple problems that are best suited for the prefrontal cortex? The best way is to ask yourself if the decision can be accurately summarized in numerical terms. For example, since most vegetable peelers are virtually identical, not much is lost when the various peelers are sorted by price. In this case, the best choice is probably the cheapest: let the rational brain take over. (Especially since the emotional brain might be misled by spiffy packaging or some other irrelevant variable.) And if someone really doesn't care about strawberry jam—he or she just wants something to put on a peanut butter sandwich—then this deliberate decision-making strategy can also be applied to jam. Or wine. Or brands of cola. Or any domain in which the details of the product aren't particularly important. In these situations, remember what we learned about expensive wine in chapter 5, and don't spend too much money on overpriced items that won't be appreciated. (After all, cheaper wines often taste better than more expensive ones in blind taste tests!) If the decision doesn't matter all that much, the prefrontal cortex should take the time to carefully assess and analyze the options.

On the other hand, for important decisions about complex items—leather couches, cars, and apartments, for example—categorizing by price alone will eliminate a lot of essential information. Perhaps the cheapest couch is of inferior quality, or maybe you don't like the way it looks. And should anyone really choose an apartment or a car based on a single variable, such as the monthly rent or the amount of horsepower? As Dijksterhuis demonstrated, when you ask the prefrontal cortex to make these sorts of decisions, it makes consistent mistakes. You'll end up with an ugly couch in the wrong apartment. It might sound ridiculous, but it makes scientific sense: Think *less* about those items that you care a lot about. Don't be afraid to let your emotions choose.

Likewise, there's a whole subset of everyday decisions—those mundane choices that don't really matter—that could benefit from a little more conscious deliberation. Too often, we let our impulses make the easy decisions for us. A person will pick a vegetable peeler, laundry detergent, or boxer shorts on a whim and automatically trust his instincts when he gets an obvious poker hand. But these are precisely the sorts of emotion-driven decisions that might benefit from rational analysis.

NOVEL PROBLEMS ALSO REQUIRE REASON. Before you entrust a mystery to the emotional brain, before deciding to let your instincts make a big bet in poker or fire a missile at a suspicious radar blip, ask yourself a question: How does your past experience help solve this particular problem? Have you played poker hands like this before? Seen blips like this before? Are these feelings rooted in experience, or are they just haphazard impulses?

If the problem really is unprecedented—if it's like a complete hydraulic failure in a Boeing 737—then emotions can't save you. Stop and think and let your working memory tackle the dilemma. The only way out of a unique mess is to come up with a creative solution, like Al Haynes did when he realized that he couldn't steer the plane in the ordinary way but that it was possible to steer the plane with the thrust levers. Such insights require the flexible neurons of the prefrontal cortex.

However, this doesn't mean that our emotional state is irrelevant. Mark Jung-Beeman, the scientist who studies the neuroscience of insight, has shown that people in good moods are significantly better at solving hard problems that require insight than people who are cranky and depressed. (Happy people solve nearly 20 percent more word puzzles than unhappy people.) He speculates that this is because the brain areas associated with executive control, such as the prefrontal cortex and the ACC, aren't as preoccupied with managing emotional life. In other words, they aren't worrying about why you're not happy, which means they are free to solve the problem at hand. The end result is that the rational brain can focus on what it needs to focus on, which is coming up with a solution for the unprecedented situation that you've found yourself in.

EMBRACE UNCERTAINTY. Hard problems rarely have easy solutions. There is no single way to win a poker hand, and there is no guaranteed path to making money in the stock market. Pretending that the mystery has been erased results in the dangerous trap of certainty. You are so confident you're right that you neglect all the evidence that contradicts your conclusion. You fail to notice that those Egyptian tanks on the border aren't merely engaging in a training exercise. Of course, there's not always time to engage in a lengthy cognitive debate. When an Iraqi missile is zooming toward you or when you're about to get crushed by a blitzing linebacker, you need to act. But whenever possible, it's essential to extend the decision-making process and properly consider the argument unfolding inside your head. Bad decisions happen when that mental debate is cut short, when an artificial consensus is imposed on the neural quarrel.

There are two simple tricks to help ensure that you never let certainty interfere with your judgment. First, always entertain competing hypotheses. When you force yourself to interpret the facts through a different, perhaps uncomfortable lens, you often discover that your beliefs rest on a rather shaky foundation. For instance, when Michael Binger is convinced that another player is bluffing, he tries to think about how the player would be acting if he *wasn't* bluffing. He is his own devil's advocate.

Second, continually remind yourself of what you *don't* know. Even the best models and theories can be undone by utterly unpredictable events. Poker players call these "bad beats," and every player has stories about the hands he lost because he got the one card he wasn't expecting. "One of the things I learned from counting cards in blackjack," Binger says, "is that even when you have an edge, and counting cards is definitely an edge, your margin is still really slim. You can't get too cocky." When you forget that you have blind spots, that you have no idea what cards the other players are holding or how they'll behave, you're setting yourself up for a nasty surprise. Colin Powell made a number of mistakes in the run-up to the Iraq war, but his advice to his intelligence officers was psychologically astute: "Tell me what you know," he told his advisers. "Then tell me what you don't know, and only then can you tell me what you think. Always keep those three separated."

YOU KNOW MORE THAN YOU KNOW. One of the enduring paradoxes of the human mind is that it doesn't know itself very well. The conscious brain is ignorant of its own underpinnings, blind to all that neural activity taking place outside the prefrontal cortex. This is why people have emotions: they are windows into the unconscious, visceral representations of all the information we process but don't perceive.

For most of human history, the emotions have been disparaged because they're so difficult to analyze—they don't come with reasons, justifications, or explanations. (As Nietzsche warned, we are often most ignorant of what is closest to us.) But now, thanks to the tools of modern neuroscience, we can see that emotions have a logic all their own. The jitters of dopamine help keep track of reality, alerting us to all those subtle patterns that we can't consciously detect. Different emotional areas evaluate different aspects of the world, so your insula naturally takes the cost of an item into account (unless you're paying with a credit card), and the NAcc automatically figures out how you feel about a certain brand of strawberry jam. The anterior cingulate monitors surprises, and the amygdala helps point out the radar blip that just doesn't look right.

The emotional brain is especially useful at helping us make hard decisions. Its massive computational power—its ability to process millions of bits of data in parallel—ensures that you can analyze all the relevant information when assessing alternatives. Mysteries are broken down into manageable chunks, which are then translated into practical feelings.

The reason these emotions are so intelligent is that they've managed to turn mistakes into educational events. You are constantly benefiting from experience, even if you're not consciously aware of the benefits. It doesn't matter if your field of expertise is backgammon or Middle East politics, golf or computer programming: the brain always learns the same way, accumulating wisdom through error.

There are no shortcuts to this painstaking process; becoming an expert just takes time and practice. But once you've developed expertise in a particular area — once you've made the requisite mistakes — it's important to trust your emotions when making decisions in that domain. It is feelings, after all, and not the prefrontal cortex, that capture the wisdom of experience. Those subtle emotions saying shoot down the radar blip, or go all in with pocket kings, or pass to Troy Brown are the output of a brain that has learned how to read a situation. It can parse the world in practical terms, so that you know what needs to be done. When you overanalyze these expert decisions, you end up like the opera star who couldn't sing.

And yet, this doesn't mean the emotional brain should always be trusted. Sometimes it can be impulsive and short-sighted. Sometimes it can be a little *too* sensitive to patterns, which is why people lose so much money playing slot machines. However, the one thing you should always be doing is considering your emotions, thinking about why you're feeling what you're feeling. In other words, act like the television executive carefully analyzing the reactions of the focus group. Even when you choose to ignore your emotions, they are still a valuable source of input.

THINK ABOUT THINKING. If you're going to take only one idea away from this book, take this one: Whenever you make a decision, be aware of the kind of decision you are making and the kind of thought process it requires. It doesn't matter if you're choosing between wide receivers or political candidates. You might be playing poker or assessing the results of a television focus group. The best way to make sure that you are using your brain properly is to study your brain at work, to listen to the argument inside your head.

Why is thinking about thinking so important? First, it helps us steer clear of stupid errors. You can't avoid loss aversion unless you know that the mind treats losses differently than gains. And you'll probably think too much about buying a house unless you know that such a strategy will lead you to buy the wrong property. The mind is full of flaws, but they can be outsmarted. Cut up your credit cards and put your retirement savings in a low-cost index fund. Prevent yourself from paying too much attention to MRI images, and remember to judge a wine before you know how much it costs. There is no secret recipe for decision-making. There is only vigilance, the commitment to avoiding those errors that can be avoided.

Of course, even the most attentive and self-aware minds will still make mistakes. Tom Brady, after the perfect season of 2008, played poorly in the Super Bowl. Michael Binger, after a long and successful day of poker, always ends up regretting one of his bets. The most accurate political experts in Tetlock's study still made plenty of inaccurate predictions. But the best decision-makers don't despair. Instead, they become students of error, determined to learn from what went wrong. They think about what they could have done differently so that the next time their neurons will know what to do. This is the most astonishing thing about the human brain: it can always improve itself. Tomorrow, we can make better decisions.

# Coda

There are certain statistics that seem like they'll never change: the high school dropout rate, the percentage of marriages that end in divorce, the prevalence of tax fraud. The same used to be true of plane crashes that were due to pilot error. Despite a long list of aviation reforms, from mandatory pilot layovers to increased classroom training, that percentage refused to budge from 1940 to 1990, holding steady at around 65 percent. It didn't matter what type of plane was being flown or where the plane was going. The brute fact remained: most aviation deaths were due to bad decisions in the cockpit.

But then, starting in the early 1990s, the percentage of crashes attributed to pilot error began to decline rapidly. According to the most current statistics, mistakes by the flight crew are responsible for less than 30 percent of all plane accidents, with a 71 percent reduction in the number of accidents caused by poor decision-making. The result is that flying has become safer than ever. According to the National Transportation Safety Board, flying on a commercial plane has a fatality rate of 0.04 per one hundred million passenger miles, making it the least dangerous form of travel by far. (In contrast, driving has a fatality rate of 0.86.) Since 2001, pilot error has caused only one fatal jetliner crash in the United States, even though more than thirty thousand flights take off every day. The most dangerous part of traveling on a commercial airplane is the drive to the airport.

What caused the dramatic reduction in pilot error? The first factor was the introduction in the mid-1980s of realistic flight simulators. For the first time, pilots could practice making decisions. They could refine their reactions to a sudden downdraft in a thunderstorm and practice landing with only one engine. They could learn what it would be like to fly without wing flaps and to land on a tarmac glazed with ice. And they could do all this without leaving the ground.

These simulators revolutionized pilot training. "The old way of teaching pilots was the 'chalk and talk' method," says Jeff Roberts, the group president of civil training at CAE, the largest manufacturer of flight simulators. Before pilots ever entered the cockpit, they were forced to sit through a long series of classroom lectures. They learned all the basic maneuvers of flight while on the ground. They were also taught how to react in the event of various worst-case scenarios. What should you do if the landing gear won't deploy? Or if the plane is struck by lightning? "The problem with this approach," Roberts says, "is that everything was abstract. The pilot has this body of knowledge, but they'd never applied it before."

The benefit of a flight simulator is that it allows pilots to internalize their new knowledge. Instead of memorizing lessons, a pilot can train the emotional brain, preparing the parts of the cortex that will actually make the decision when up in the air. As a result, pilots who are confronted with a potential catastrophe during a real flight—like an engine fire in the air above Tokyo—already know what to do. They don't have to waste critical moments trying to remember what they learned in the classroom. "A plane is traveling four hundred miles per hour," Roberts says. "It's the rare emergency when you've got time to think about what your flight instructor told you. You've got to make the right decision right away."

Simulators also take advantage of the way the brain learns from experience. After pilots complete their "flight," they are forced to endure an exhaustive debriefing. The instructor scrutinizes all of their decisions, so that the pilots think about why, exactly, they decided to gain altitude after the engine fire, or why they chose to land in the hailstorm. "We want pilots to make mistakes in the simulator," Roberts says. "The goal is to learn from those mistakes when they don't count, so that when it really matters, you can make the right decision." This approach targets the dopamine system, which improves itself by studying its errors. As a result, pilots develop accurate sets of flight instincts. Their brains have been prepared in advance.

There was one other crucial factor in the dramatic decline of pilot error: the development of a decision-making strategy known as Cockpit Resource Management (CRM). The impetus for CRM came from a large NASA study in the 1970s of pilot error; it concluded that many cockpit mistakes were attributable, at least in part, to the "God-like certainty" of the pilot in command. If other crew members had been consulted, or if the pilot had considered other alternatives, then some of the bad decisions might have been avoided. As a result, the goal of CRM was to create an environment in which a diversity of viewpoints was freely shared.

Unfortunately, it took a tragic crash in the winter of 1978 for airlines to decide to implement this new system. United Flight 173 was a crowded DC-8 bound for Portland, Oregon. About ten miles from the runway, the pilot lowered the landing gear. He noticed that two of his landing-gear indicator lights remained off, suggesting that the front wheels weren't properly deployed. The plane circled around the airport while the crew investigated the problem. New bulbs were put in the dashboard. The auto-pilot computers were reset. The fuse box was double-checked. But the landing-gear lights still wouldn't turn on.

The plane circled for so long that it began to run out of fuel. Unfortunately, the pilot was too preoccupied with the landing gear to notice. He even ignored the flight engineer's warning about the fuel levels. (One investigator described the pilot as "an arrogant S.O.B.") By the time the pilot looked at his gas gauge, the engines were beginning to shut down. It was too late to save the plane. The DC-8 crash-landed in a sparsely populated Portland suburb, killing ten and seriously wounding twenty-four of the 189 on board. Crash investigators later concluded that there was no problem with the landing gear. The wheels were all properly deployed; it was just a faulty circuit.

After the crash, United trained all of its employees with CRM. The captain was no longer the dictator of the plane. Instead, flight crews were expected to work together and constantly communicate with one another. Everyone was responsible for catching errors. If fuel levels were running low, then it was the job of the flight engineer to make sure the pilot grasped the severity of the situation. If the copilot was convinced that the captain was making a bad decision, then he was obligated to dissent. Flying a plane is an extremely complicated task, and it's essential to make use of every possible resource. The best decisions emerge when a multiplicity of viewpoints are brought to bear on the situation. The wisdom of crowds also applies in the cockpit.

Remember United Flight 232, which lost all hydraulic power? After the crash-landing, the pilots all credited CRM with helping them make the runway. "For most of my career, we kind of worked on the concept that the captain was *the* authority on the aircraft," says Al Haynes, the captain of Flight 232. "And we lost a few airplanes because of that. Sometimes the captain isn't as smart as we thought he was." Haynes freely admits that he couldn't have saved the plane by himself that day. "We had 103 years of flying experience there in the cockpit [on Flight 232], trying to get that airplane on the ground. If I hadn't used CRM, if we had not had everybody's input, it's a cinch we wouldn't have made it."

In recent years, CRM has moved beyond the cockpit. Many hospitals have realized that the same decision-making techniques that can prevent pilot error can also prevent unnecessary mistakes during surgery. Consider the experience of the Nebraska Medical Center, which began training its surgical teams in CRM in 2005. (To date, more than a thousand hospital employees have undergone the training.) The mantra of the CRM program is "See it, say it, fix it"; all surgical-team members are encouraged to express their concerns freely to the attending surgeon. In addition, team members engage in postoperation debriefings at which everyone involved is supposed to share his or her view of the surgery. What mistakes were made? And how can they be avoided the next time?

The results at the Nebraska Medical Center have been impressive. A 2007 analysis found that after fewer than six months of CRM training, the percentage of staff members who "felt free to question the decisions of those with more authority" had gone from 29 percent to 86 percent. More important, this increased willingness to point out potential errors led to a dramatic decrease in medical mistakes. Before CRM training, only around 21 percent of all cardiac surgeries and cardiac catheterizations were classified as "uneventful cases," meaning that nothing had gone wrong. After CRM training, however, the number of "uneventful cases" rose to 62 percent.

The reason CRM is so effective is that it encourages flight crews and surgical teams to think together. It deters certainty and stimulates debate. In this sense, CRM creates the ideal atmosphere for good decision-making, in which a diversity of opinions is openly shared. The evidence is looked at from multiple angles, and new alternatives are considered. Such a process not only prevents mistakes but also leads to startling new insights.

TO SIT IN a modern airplane cockpit is to be surrounded by computers. Just above the windshield are the autopilot terminals, which can keep a plane on course without any input from the pilot. Right in front of the thrust levers is a screen relaying information about the state of the plane, from its fuel levels to the hydraulic pressure. Nearby is the computer that monitors the flight path and records the position and speed of the plane. Then there's the GPS panel, a screen for weather updates, and a radar monitor. Sitting in the captain's chair, you can tell why it's called the glass cockpit: everywhere you look there's another glass screen, the digital output of the computers underneath.

These computers are like the emotional brain of the plane. They process a vast amount of information and translate that information into a form that can be quickly grasped by the pilot. The computers are also redundant, so every plane actually contains multiple autopilot systems running on different computers and composed in different programming languages. Such diversity helps prevent mistakes, since each system is constantly checking itself against the other systems.

These computers are so reliable that they perform many of their tasks without any pilot input. If, for example, the autopilot senses a strong headwind, it will instantly increase thrust in order to maintain speed. The pressure in the cabin is seamlessly adjusted to reflect the altitude of the plane. If a pilot is flying too close to another plane, the onboard computers emit loud warning sirens, forcing the flight crew to notice the danger. It's as if the plane has an amygdala.

Pilots are like the plane's prefrontal cortex. Their job is to monitor these onboard computers, to pay close attention to the data on the cockpit screens. If something goes wrong, or if there's a disagreement among the various computers, then it's the responsibility of the flight crew to resolve the problem. The pilots must immediately intervene and, if necessary, take control of the plane. The pilots must also set the headings, supervise the progress of the flight, and deal with the inevitable headaches imposed by air-traffic control. "People who aren't pilots tend to think that when the autopilot is turned on, the pilot can just take a nap," my flight instructor in the simulator says. "But planes don't fly themselves. You can't ever relax in the cockpit. You always have to be watching, making sure everything is going according to plan."

Consider the cautionary tale of a crowded Boeing 747 traveling from Miami to London in May 2000. The runway at Heathrow was shrouded in dense fog, so the pilots decided to make an automated landing, or what's known as a category IIIc approach. During the initial descent, all three autopilot systems were turned on. However, when the plane reached an altitude of a thousand feet, the lead autopilot system suddenly shut down for no apparent reason. The pilots decided to continue with the approach, since the 747 is designed to be able to make automated landings with only two autopilot systems. The descent went smoothly until the plane was fifty feet above the runway, or about four seconds from touchdown. At that point, the autopilot abruptly tilted the nose of the plane downward, so that its rate of descent was four times faster than normal. (Investigators would later blame a programming error for the mistake.) The pilot quickly intervened and yanked back on the control column so that the plane wouldn't hit the runway nose first. The landing was still rough — the plane suffered some minor structural damage — but the quick reactions of the flight crew prevented a catastrophe.

Events like this are disturbingly common. Even redundant autopilot systems will make mistakes. They'll disengage or freeze or steer the plane in dangerous ways. Unless a pilot is there to correct the error, to turn off the computer and pull up the nose, the plane will fly itself into the ground.

Of course, pilots aren't perfect either. They sometimes fail to notice when they're getting too close to another plane, or they struggle to monitor all the different gauges in the cockpit. In fact, if pilots had to rely on their own instincts, they wouldn't even be able to fly through clouds. (The inner ear can't detect blind turns, which means that it's very tough to fly straight without proper instruments or visual cues.) Then there are the pilots that micromanage the flight—constantly overruling the autopilot or fiddling with the path of the plane. They dramatically increase the likelihood of human error, acting like people who rely too heavily on their prefrontal cortices.

When the onboard computers and pilot properly interact, it's an ideal model for decision-making. The rational brain (the pilot) and the emotional brain (the cockpit computers) exist in perfect equilibrium, each system focusing on those areas in which it has a comparative advantage. The reason planes are so safe, even though both the pilot and the autopilot are fallible, is that both systems are constantly working to correct each other. Mistakes are fixed before they spiral out of control.

The payoff has been huge. "Aviation is just about the only field that consistently manages to operate at the highest level of performance, which is defined by six sigma," Roberts says, using the managerial buzzword for any process that produces fewer than 3.4 defects per one million opportunities. "Catastrophic error in planes is incredibly, incredibly rare. If it wasn't, nobody would ever get on board. The fact of the matter is that the aviation industry needs to be perfect, and so we found ways to be as close to perfect as humanly possible."

The safety of flight is a testament to the possibility of improvement. The reduction in the pilot-error rate is a powerful reminder that mistakes are not inevitable, that planes don't have to crash. As the modern cockpit demonstrates, a few simple innovations and a little self-awareness can dramatically improve the way people think, so that both brain systems are used in their ideal context. The aviation industry took decision-making seriously—they made a science of pilot error—and the result has been a stunning advance in performance.

The first step to making better decisions is to see ourselves as we really are, to look inside the black box of the human brain. We need to honestly assess our flaws and talents, our strengths and shortcomings. For the first time, such a vision is possible. We finally have tools that can pierce the mystery of the mind, revealing the intricate machinery that shapes our behavior. Now we need to put this knowledge to work.

*Acknowledgments*

*Notes*

*Bibliography*

# *Acknowledgments*

I decided to write a book about decision-making because I couldn't make a decision about Cheerios. There I was, aimlessly wandering the cereal aisle of the supermarket, trying to choose between the Apple Cinnamon and Honey Nut varieties. It was an embarrassing waste of an afternoon, and yet it happened to me *all the time*. Eventually, I decided that enough was enough: I was going to understand what was happening inside my brain as I contemplated my breakfast options. So thank you, General Mills, for making so many different kinds of Cheerios.

There is, of course, a lot of work that has to be done before the sudden epiphany ("I should write a book about decisions!") actually becomes a book. Once again, Amanda Cook, my editor at Houghton Mifflin Harcourt, was a godsend. She took a messy, convoluted, unstructured manuscript and managed to find the thread that brought it all together. She suggested stories, fixed my prose, and talked me through my confusion. She is the kind of editor—thoughtful, whip-smart, and generous—that every writer dreams about. I am so fortunate to have her on my side.

I also benefited from the extremely helpful comments and suggestions of many friends. Robert Krulwich, as always, taught me how to tell a story; the team at Radio Lab, including Jad Abumrad, Lulu Miller, and Soren Wheeler, helped me figure out which stories were worth telling. Gareth Cook edited bits and pieces of this book for the *Boston Globe* Ideas section, while Laura McNeil and Adam Bly allowed me to explore the field of neuroscience in *Seed* magazine. David Pook taught me about moral philosophy, while Ted Trimmer made sure I got my aviation facts right.

I'm also incredibly grateful to all of the scientists who took the time to talk with a curious writer. I asked them all sorts of naïve questions and mispronounced any number of brain regions, and yet, without exception, they've been understanding, patient, and helpful. The mistakes that remain are my own fault. And then there are the people, like Ann Klinestiver, Al Haynes, Herb Stein, Ralph Jimenez, Felice Belman, Mike Pride, and Michael Binger, who let me share their personal stories in these pages. It's been an honor.

I probably wouldn't be a writer if my wonderful agent, Emma Parry, hadn't spotted an article I wrote years ago for *Seed* and convinced me that it could become an improbable book called *Proust Was a Neuroscientist*. She's a consummate problem-solver and endless source of good ideas. I'm so grateful to her and everyone at Fletcher and Parry, especially Christy Fletcher and Melissa Chinchillo. Nick Davies, my editor at Canongate Books, has been a joy to work with. He also managed to teach me the rules of cricket, for which he deserves some sort of very prestigious award. Tracy Roe, the world's finest copyeditor, has saved me yet again from an embarrassing number of errors and flabby sentences.

And then there is my family: my sister Rachel is not only a fantastic modern dancer, she also gave me great advice on the manuscript; my brother, Eli, helped me think about the real-world implications of these scientific theories (and kept me well supplied with music mixes). I even benefited from the red pen of my grandmother Louise. My father was a patient listener and a fount of relevant factoids and articles, e-mailed to me daily. My mother was an essential reader; I don't know where she finds the time to read my first drafts, but I can't imagine writing without her feedback.

My girlfriend, Sarah Liebowitz, has read this book dozens of times (I'm not exaggerating) in all of its various drafts and guises. This book wouldn't exist without her insightful criticisms, cheerful support, and love. By the time you read this, Sarah will be my wife, which is, without a doubt, the best decision I've ever made.

# *Notes*

## INTRODUCTION

## 1. THE QUARTERBACK IN THE POCKET

14 *"Elliot emerged as"*: Damasio, *Descartes' Error*, 43.
   *"He was always"*: Ibid., 44.
16 *"I suggested two"*: Ibid., 193.
19 *"Many of these"*: Herb Stein, telephone interview with the author, October 15, 2007.
25 *But people can't*: Gordon, ed., *Your Brain on Cubs*, chapter 3.
   *For instance, a study*: Muller et al., "World-Class Cricket Batsmen," 2162–86.
26 *"The facts of"*: James, *Principles of Psychology*, 389.

## 2. THE PREDICTIONS OF DOPAMINE

32 *"It was ours"*: Klein, *Sources of Power*, 36–38; Finlan, *The Royal Navy*, 147–50.
34 *In 1954*: Olds and Milner, "Positive Reinforcement," 419–27.
37 *The monkey felt*: Schultz et al., "Predictive Reward Signal"; Hollerman and Schultz, "Dopamine Neurons."
38 *This fast cellular*: P. Gaspar et al., "Catecholamine Innervation."
39 *Researchers at Oxford*: Kennerley, "Optimal Decision-Making."
   *If the monkey*: Hayden and Platt, "Fool Me Once."
40 *Because they have*: Klein et al., "Genetically Determined Differences."
41 *"You're probably"*: Zweig, *Your Money and Your Brain*, 69.
   *Instead, human emotions*: Cohen et al., "Reinforcement Learning."
42 *The roots of schizophrenia*: Juckel et al., "Dysfunction of Ventral Striatal"; Dolan et al., "Dopaminergic Modulation."
   *Most antipsychotic medications*: Adams and Moghaddam, "Corticolimbic Dopamine Neurotransmission."
44 *Tesauro set out*: Montague, *Why Choose This Book?*, 93–100.
45 *"Anytime you've got"*: Read Montague, interview with the author, May 7, 2007.
46 *They use their*: Ito, "Performance Monitoring."
47 *The subject's feelings figured*: Bechara et al., "Deciding Advantageously."
   *By playing the*: Oya et al., "Electrophysiological Correlates."
48 *The investments that*: Betsch et al., "Different Principles."
49 *"He wanted the"*: Bill Robertie, telephone interview with the author, October 10, 2007.
51 *"a person who"*: Mackay, *A Dictionary of Scientific Quotations*, 35.
52 *The other students*: Cimpian et al., "Subtle Linguistic Cues"; Bronson, "How Not to Talk"; Mangels et al., "Why Do Beliefs."

## 3. FOOLED BY A FEELING

57 *"I lost control"*: Ann Klinestiver, telephone interview with the author, July 10, 2007.

58 *"You can feel"*: Ibid., July 21, 2007.

59 *Since slot machines*: Cooper, "Sit and Spin."

60 *According to Wolfram*: Wolfram Schultz, telephone interview with the author, August 14, 2007.

63 *There was absolutely*: Gilovich, Vallone, and Tversky, "The Hot Hand."

66 *"We made it"*: Mlodinow, *Drunkard's Walk*, 175.
  *"So he makes"*: Gilovich, *How We Know*, 17.

67 *Fama looked at*: Fama, "Random Walks."
  *One of Montague's*: Lohrenz et al., "Neural Signature."

70 *The investor who does*: Zweig, *Your Money and Your Brain*, 226.

72 *Take Nondumiso Sainsbury*: *Deal or No Deal*, NBC, air date September 18, 2006.

74 *An exhaustive analysis*: Post et al., "Deal or No Deal?"

76 *This mental defect*: Kahneman and Tversky, "Prospect Theory."

77 *"confused about the"*: T. MaCurdy and J. Shoven, "Accumulating Pension Wealth with Stocks and Bonds," working paper, Stanford University, cited in Benartzi and Thaler, "Myopic Loss Aversion."

78 *Markowitz was so*: Zweig, *Your Money and Your Brain*, 4.
  *Loss aversion also*: Odean, "Are Investors Reluctant."

79 *Consider this experiment*: Shiv et al., "Investment Behavior."

82 *"Every day, I see"*: Herman Palmer, telephone interview with the author, October 19, 2007, and personal interview with the author, Bronx, NY, January 11, 2008.

86 *They no longer*: Prelec and Simester, "Always Leave Home Without It."

88 *In fact, subprime*: Brooks and Simon, "Subprime Debacle."

89 *Jonathan Cohen, a neuroscientist*: McClure et al., "Separate Neural Systems."

90 *This also helps*: Wilson and Daly, "Do Pretty Women."
  *"Our emotional brain"*: Bradt, "Brain Takes Itself On."
  *Lawrence Ausubel, an economist*: Ausubel, "Adverse Selection."

91 *"Our emotions are like"*: George Loewenstein, telephone interview with the author, March 10, 2007.
  *Shlomo Benartzi and Richard Thaler*: Benartzi and Thaler, "Save More Tomorrow."

## 4. The Uses of Reason

94   *"It sounds like"*: Maclean, *Young Men and Fire*, 35.
98   *"To hell with"*: Ibid., 95.
99   *Although the people*: Weltman, "Perceptual Narrowing."
     *"It just seemed"*: Maclean, *Young Men and Fire*, 100.
100  *Pressed tight against*: Schoenemann, "Prefrontal White Matter."
     *But he still*: Wynn and Coolidge, "The Expert Neandertal Mind."
102  *Consider the case*: Heilman, *Matter of Mind*, 81.
104  *"You see this"*: Kenneth Heilman, interview with the author, June 28, 2007.
106  *When neuroscientists used*: De Martino et al., "Frames, Biases."
108  *"Anyone can become"*: Aristotle, *Nicomachean Ethics*, 48.
     *The CBS method*: Napoli, *Audience Economics*, 41.
109  *"Quantitative data"*: Brian Graden, interview on *Frontline*.
     *"The shows weren't"*: Television executive (who asked to remain anonymous), telephone interview, November 12, 2007.
112  *Mischel's results were*: Shoda, Mischel, and Peake, "Predicting Adolescent Cognitive."
113  *In November 2007*: Shaw et al., "Attention-Deficit/Hyperactivity Disorder."
114  *A recent study*: Galvan et al., "Earlier Development of the Accumbens."
115  *The most important*: MacDonald et al., "Dissociating the Role."
116  *"No other brain"*: Earl Miller, interview with the author, April 1, 2008.
117  *In 2007, in a paper*: Buschman and Miller, "Top-Down versus Bottom-Up."
     *The end result*: Wood and Grafman, "Human Prefrontal Cortex."
118  *Less than 20 percent*: Fleck and Weisberg, "The Use of Verbal Protocols."
     *The subject repeats*: Davidson and Sternberg, eds., *Psychology of Problem Solving*, 151.
     *Instead of trying*: Colvin et al., "The Effects of Frontal Lobe Lesions."
119  *The first brain*: Kounios et al., "The Prepared Mind."
     *"You're getting rid"*: Mark Jung-Beeman, interview with the author, February 15, 2008.
120  *He'd flown this*: Captain Al Haynes, interview with the author, January 21, 2008.
122  *"It was an"*: Dennis Fitch, interview, *Seconds from Disaster*, National Geographic Channel.

127 *"The performance was":* http://amelia.db.erau.edu/reports/ntsb/aar/
AAR90-06.pdf.

128 *By analyzing the:* as cited in Homer-Dixon, *The Ingenuity Gap*,
18–20.

129 *According to Predmore:* Predmore, "Dynamics of Group Performance"; Helmreich, "Managing Human Error in Aviation."

130 *For instance, studies:* Kandel et al., *Principles of Neural Science*,
359.
*Scientists refer to:* Sandkuhler and Bhattacharya, "Deconstructing
Insight."

131 *Numerous studies have:* Colom et al., "Working Memory."

## 5. Choking on Thought

134 *"It caught me":* Fleming, *The Inner Voice*, 100.

135 *"I had been":* Ibid., 80.

138 *By concentrating on:* Beilock and Carr, "On the Fragility"; Beilock
et al., "When Paying Attention."

139 *Choking is merely:* Gucciardi and Dimmock, "Choking under Pres-
sure."
*Claude Steele, a professor:* Steele and Aronson, "Stereotype Threat";
Steele, "Thin Ice."

140 *"What you tend":* Gladwell, "The Art of Failure."

141 *A few years:* Wilson and Schooler, "Thinking Too Much."

143 *So Wilson came:* Wilson et al., "Introspecting About Reasons."

144 *As Ap Dijksterhuis:* Dijksterhuis et al., "The Rational Unconscious."

146 *Because people* expected: Wager et al., "Placebo-Induced Changes."
*Look, for example:* Shiv, Carmon, and Ariely, "Placebo Effects of
Marketing Actions."

147 *"We have these":* Baba Shiv, telephone interview with the author,
February 20, 2008.
*Researchers at Caltech:* Plassmann et al., "Marketing Actions."

148 *"We don't realize":* Antonio Rangel, telephone interview with the
author, February 20, 2008.

150 *"There seems to":* Miller, "The Magical Number Seven."
*Consider this experiment:* Shiv and Fedorikhin, "Heart and Mind."

152 *A bad mood:* Gailliot et al., "Self-Control Relies on Glucose."

153 *But just as:* Geier, Rozin, and Doros, "Unit Bias."
*The group with:* Wansink, *Mindless Eating.*
*Richard Thaler, an:* Thaler, *The Winner's Curse*, 107–22.

155 *As Thaler notes:* Bennett, "When Shove Comes to Push."

156  *A few years:* Ariely, Loewenstein, and Prelec, "Coherent Arbitrariness"; "Tom Sawyer and the Construction of Value."

158  *"You know you're":* Dan Ariely, interview with the author, May 29, 2008.

159  *While the extra:* Grove et al., "Clinical versus Mechanical Prediction."

160  *Back pain is:* Lehrer, "Psychology of Back Pain."

162  *In a 1994 study:* Jensen et al., "Magnetic Resonance Imaging."

163  *"A lot of":* Sean Mackey, interview with the author, June 1, 2007.
     *A large study:* Jarvik et al., "Rapid Magnetic Resonance Imaging."

165  *They wanted to:* Deyo, Nachemson, and Mirza, "Spinal-Fusion Surgery."
     *"What's going on":* John Sarno, interview with the author, June 1, 2007.

## 6. THE MORAL MIND

167  *Sadism was entertaining:* Sullivan and Maiken, *Killer Clown.*

169  *"[Gacy] appears to have":* Wilkinson, "Conversations with a Killer."

170  *The acts of:* Levenson, Carstensen, and Gottman, "The Influence of Age and Gender."
     *"You know when":* James Blair, telephone interview with the author, April 4, 2007.

171  *The emotional areas:* Deeley et al., "Facial Emotion Processing."
     *The main problem:* Blair, Mitchell, and Blair, *The Psychopath.*
     *Brain-imaging studies:* Berthoz et al., "Affective Response."

172  *"Moral judgment is":* Haidt, *Happiness Hypothesis,* 22.

173  *Consider this moral:* Haidt, "The Emotional Dog."

174  *"What happens in":* Sommers, "Jonathan Haidt."

176  *Consider this elegant:* Greene et al., "An fMRI Investigation."

178  *"Our primate ancestors":* Joshua Greene, telephone interview with the author, June 25, 2007.

179  *"It is fear":* Marshall, *Men Against Fire,* 78.
     *"At the most":* Ibid., 79.
     *"What is being":* Grossman, *On Combat,* 254.

181  *However, the researchers:* Oosterbeek, Sloof, and van de Kuilen, "Differences in Ultimatum Game Experiments"; Henrich et al., *Foundations of Human Sociality."*

182  *"Seen through the":* Frank, *Passions Within Reason,* ix.
     *"As we have":* Smith, *Theory of Moral Sentiments,* 3–4.
     *The reason a proposer:* Sanfey et al., "The Neural Basis."

183 *Because they intensely:* Tankersley, Stowe, and Huettel, "Altruism Is Associated."

184 *When they chose:* Moll et al., "Human Fronto-mesolimbic Networks."

185 *It now appears:* Dapretto et al., "Understanding Emotions."
*"They allow us":* Blakeslee, "Cells That Read Minds."

186 *A face generated:* Schultz et al., "The Role of the Fusiform Face Area."

187 *Instead of giving:* Hoffman, McCabe, and Smith, "Social Distance and Other-Regarding Behavior."
*"The experience of":* Keltner, "Power Paradox."
*Paul Slovic, a:* Small, Loewenstein, and Slovic, "Sympathy and Callousness."

191 *"If monkeys have":* Blum, *Love at Goon Park,* 231.

192 *One Canadian couple:* Stout, *The Sociopath Next Door,* 132.

193 *When neuroscientists imaged:* Chugani et al., "Local Brain Functional Activity."
*The orphans also:* Parker et al., "The Impact of Early Institutional Rearing."
*Finally, the neglected:* Carter, "The Chemistry of Child Neglect."
*In the early 1980s:* George and Main, "Social Interactions"; Main and George, "Responses of Abused and Disadvantaged Toddlers."

194 *They were simply:* Taylor et al., "Neural Bases of Regulatory Deficits."

195 *Consider this poignant:* Masserman, Wechkin, and Terris, "Altruistic Behavior in Rhesus Monkeys."

## 7. THE BRAIN IS AN ARGUMENT

196 *"There were a":* Ralph Jimenez, at *Monitor* editors' interview with the author, March 25, 2008.

197 *"I was still":* Felice Belman, ibid.
*"The candidates are":* Mike Pride, ibid.

199 *"The point is":* Antoine Bechara, interview, *Radio Lab,* May 12, 2008.

200 *Consider this clever:* Knutson et al., "Neural Predictors."

202 *In fact, researchers:* Inman, McAlister, and Hoyer, "Promotion Signal."

204 *Although it was:* Green, Palmquist, and Schickler, *Partisan Hearts,* 2.
*Drew Westen, a:* Westen et al., "The Neural Basis of Motivated Reasoning."

205 *"Essentially, it appears":* Westen, *The Political Brain,* 17.

206 *"Voters think that"*: Achen and Bartels, "It Feels Like We're Thinking."
     *Let's look at:* Brock and Balloun, "Behavioral Receptivity."
207 *In 1984, the:* Tetlock, *Expert Political Judgment.*
209 *"The dominant danger":* Ibid., 23.
211 *"Oh, that's easy":* Gazzaniga, *The Social Brain,* 72.
     *During the last:* Rabinovich, *The Yom Kippur War.*
212 *He persuasively argued:* Black and Morris, *Israel's Secret Wars,* 280–99; Raviv and Melman, *Every Spy a Prince,* 110–25.
216 *"The need for":* Bar-Joseph and Kruglanski, "Intelligence Failure"; Bar-Joseph, *The Watchman Fell Asleep.*
218 *He intentionally filled:* Goodwin, *Team of Rivals,* 126.

## 8. THE POKER HAND

220 *"You get so wired":* Michael Binger, interviews with the author, May 12, 2007; July 1, 2007; September 22, 2007.
232 *To answer these:* Dijksterhuis et al., "On Making the Right Choice."
236 *"Imagine being at":* Dijksterhuis and van Olden, "On the Benefits of Thinking Unconsciously."
     *"The moral of":* Dijksterhuis, "Breakthrough Ideas."
237 *The hardest calls:* Dijksterhuis and Nordgren, "A Theory of Unconscious Thought,"
     *That's why the:* Klein, "In the Digital Age."
239 *A few years:* Lo and Repin, "The Psychophysiology of Real-Time Financial Risk Processing."
240 *"One of the":* See http://alum.mit.edu/ne/opendoor/200509/lo.html.
243 *"We need to cultivate":* Tetlock, *Expert Political Judgment,* 24.

## CODA

251 *But then, starting:* Baker et al., "Pilot Error in Air Carrier Mishaps."
252 *"The old way":* Jeff Roberts, telephone interview with the author, July 12, 2007.
254 *"For most of":* Al Haynes, interview with the author, January 21, 2008.

# Bibliography

Abbot, L. F. "Balancing Homeostasis and Learning in Neural Circuits." *Zoology* 106 (2003): 365–71.

Achen, Christopher, and Larry Bartels. "It Feels Like We're Thinking: The Rationalizing Voter and Electoral Democracy." Working paper, http://www.princeton.edu/~/bartels/thinking.pdf.

Adams, Barbara, and Bita Moghaddam. "Corticolimbic Dopamine Neurotransmission Is Temporally Dissociated from the Cognitive and Locomotor Effects of Phencyclidine." *Journal of Neuroscience* 18 (1998): 5545–54.

Adolphs, R., D. Tranel, M. Koenigs, and A. Damasio. "Preferring One Taste Over Another Without Recognizing Either." *Nature Neuroscience* 8 (2005): 860–61.

Alvarez, A. *The Biggest Game in Town.* San Francisco: First Chronicle, 1983.

Anderson, S. W., Antoine Bechara, Hanna Damasio, Daniel Tranel, and Antonio R. Damasio. "Impairment of Social and Moral Behavior Related to Early Damage in Human Prefrontal Cortex." *Nature Neuroscience* 2 (1999): 1032–37.

Ariely, Dan. *Predictably Irrational.* New York: Harper, 2008.

Ariely, Dan, George Lowenstein, and Drazen Prelec. "Coherent Arbitrariness: Stable Demand Curves without Stable Preferences." *Quarterly Journal of Economics* 118 (February 2003): 73–105.

———. "Tom Sawyer and the Construction of Value." *Journal of Economic Behavior and Organization* 60 (2006): 1–10.

Aristotle. *The Nicomachean Ethics.* New York: Penguin Classics, 1980.

Ausubel, Lawrence. "Adverse Selection in the Credit Card Market." Working paper, Department of Economics, University of Maryland, June 17, 1999.

Baker, Susan, et al. "Pilot Error in Air Carrier Mishaps: Longitudinal Trends Among 558 Reports, 1983–2002." *Aviation, Space and Environmental Medicine,* January 2008.

Bar-Joseph, Uri. *The Watchman Fell Asleep.* New York: SUNY Press, 2005.

Bar-Joseph, Uri, and Arie Kruglanski. "Intelligence Failure and Need for Cognitive Closure: On the Psychology of the Yom Kippur Surprise." *Political Psychology* 24 (2003): 75–99.

Bayer, H. M., and P. Glimcher. "Midbrain Dopamine Neurons Encode a Quantitative Reward Prediction Error Signal." *Neuron* 47 (2005): 129–41.

Bechara, Antoine, Hanna Damasio, Daniel Tranel, and Antonio R. Damasio. "Deciding Advantageously Before Knowing the Advantageous Strategy." *Science* 275: 1293–96.

———. "The Iowa Gambling Task and the Somatic Marker Hypothesis." *Trends in Cognitive Science* 9 (2005): 159–62.

Beilock, S. L., and T. H. Carr. "On the Fragility of Skilled Performance: What Governs Choking under Pressure?" *Journal of Experimental Psychology: General* 130 (2001): 701–25.

Beilock, S. L., T. H. Carr, C. MacMahon, and J. L. Starkes. "When Paying Attention Becomes Counterproductive: Impact of Divided versus Skill-Focused Attention on Novice and Experienced Performance of Sensorimotor Skills." *Journal of Experimental Psychology: Applied* 8 (2002): 6–16.

Benartzi, Shlomo, and Richard Thaler. "Myopic Loss Aversion and the Equity Premium Puzzle." *Quarterly Journal of Economics* 110 (1995): 73–92.

———. "Save More Tomorrow: Using Behavioral Economics to Increase Employee Saving." *Journal of Political Economy* 112 (2004): 164–87.

Bennett, Drake. "When Shove Comes to Push." *Boston Globe,* March 2, 2008.

Berlin, Isaiah. *The Proper Study of Mankind.* New York: Farrar, Straus and Giroux, 2000.

———. *The Crooked Timber of Humanity.* Princeton: Princeton University Press, 1998.

Berthoz, S., J. Grezes, J. L. Armony, R. E. Passingham, and R. J. Dolan. "Affective Response to One's Own Moral Violations." *NeuroImage* 31 (2006): 945–50.

Betsch, Tilmann, Martina Kaufmann, Frank Lindow, Henning Plessner, and Katja Hoffmann. "Different Principles of Information Integration in Implicit and Explicit Attitude Formation." *European Journal of Social Psychology* 36 (2006): 887–905.

Black, Ian, and Benny Morris. *Israel's Secret Wars: A History of Israel's Intelligence Services.* New York: Grove, 1992.

Blair, James, Derek Mitchell, and Karina Blair. *The Psychopath: Emotion and the Brain.* New York: Wiley, 2005.

Blakeslee, Sandra. "Cells That Read Minds." *New York Times,* January 10, 2006.

Blum, Deborah. *Love at Goon Park.* New York: Perseus, 2002.

Bradt, Steve. "Brain Takes Itself On over Immediate vs. Delayed Gratification." *Harvard University Gazette,* October 21, 2004, http://www.hno.harvard.edu/gazette/2004/10.21/07-brainbattle.html.

Brock, T. C., and J. C. Balloun. "Behavioral Receptivity to Dissonant Information." *Journal of Personality and Social Psychology* 6 (1967): 413–28.

Bronson, Po. "How Not to Talk to Your Kids." *New York,* February 12, 2007.

Brooks, Rick, and Ruth Simon. "Subprime Debacle Traps Even Very Credit-Worthy." *Wall Street Journal,* December 3, 2007.

Brosnan, Sarah, and Frans de Waal. "Monkeys Reject Unequal Pay." *Nature* 425 (2003): 297–99.

Buschman, Timothy J., and Earl K. Miller. "Top-Down versus Bottom-Up Control of Attention in the Prefrontal and Posterior Parietal Cortices." *Science* 315 (2007): 1860–62.

Camerer, Colin, and Ernst Fehr. "When Does Economic Man Dominate Social Behavior?" *Science* 311 (2006): 47–52.

Carter, C. Sue. "The Chemistry of Child Neglect." *Proceedings of the National Academy of Sciences* 102 (2005): 18247–48.

Chiu, P. H., Terry Lohrenz, and P. Read Montague. "Smokers' Brains Compute, but Ignore, a Fictive Error Signal in a Sequential Investment Task." *Nature Neuroscience* 11 (2008): 514–20.

Chugani, H., Michael Behen, Otto Muzik, Csaba Juhász, Ferenc Nagy, and Diane C. Chugani. "Local Brain Functional Activity Following Early Deprivation: A Study of Postinstitutionalized Romanian Orphans." *NeuroImage* 14 (2001): 1290–1301.

Cimpian, Andrei, Holly-Marie Arce, Ellen Markman, and Carl Dweck. "Subtle Linguistic Cues Affect Children's Motivation." *Psychological Science* 18 (2007): 314–16.

Cohen, Jonathan. "The Vulcanization of the Human Brain: A Neural Perspective on Interactions between Cognition and Emotion." *Journal of Economic Perspectives* 19 (2005): 3–24.

Cohen, Jonathan, Todd Braver, and Joshua Brown. "Computational Perspectives on Dopamine Function in Prefrontal Cortex." *Current Opinion in Neurobiology* 12 (2002): 223–29.

Cohen, Michael, et al. "Reinforcement Learning Signals Predict Future Decisions." *Journal of Neuroscience* 27 (2007): 371–78.

Colom, Roberto, Carmen Flores-Mendoza, M. Ángeles Quiroga, and Jesús Privado. "Working Memory Is (Almost) Perfectly Predicted by *G*." *Intelligence* 32 (2004): 277–96.

Cooper, Marc. "Sit and Spin." *Atlantic Monthly,* December 2005.

Damasio, Antonio. *Descartes' Error.* New York: Penguin, 1995.

Damasio, Antonio, et al. "Subcortical and Cortical Brain Activity during the Feeling of Self-generated Emotions." *Nature Neuroscience* 3 (2000): 1049–56.

Dapretto, M., et al. "Understanding Emotions in Others: Mirror Neuron Dysfunction in Children with Autism Spectrum Disorders." *Nature Neuroscience* 9 (2006): 28–31.

Darwin, Charles. *The Expression of the Emotions in Man and Animals.* New York: Oxford University Press, 1998. First published in London by John Murray, 1872.

Davidson, Janet, and Robert Sternberg, eds. *The Psychology of Problem Solving.* New York: Cambridge University Press, 2003.

Dayan, Peter, and L. F. Abbot. *Theoretical Neuroscience.* Cambridge: MIT Press, 2001.

Dayan, Peter, S. Kakade, and P. R. Montague. "Learning and Selective Attention." *Nature Neuroscience* 3 (2000): 1218–23.

Deeley, Quinton, et al. "Facial Emotion Processing in Criminal Psychopathy." *British Journal of Psychiatry* 189 (2006): 533–39.

de Martino, Benedetto, Dharshan Kumaran, Ben Seymour, and Raymond J. Dolan. "Frames, Biases and Rational Decision-Making in the Human Brain." *Science* 313 (2006): 684–87.

Deyo, Richard, A. Nachemson, and S. K. Mirza. "Spinal-Fusion Surgery: The Case for Restraint." *New England Journal of Medicine* 350 (2004): 722–26.

Dijksterhuis, Ap. "Breakthrough Ideas." *Harvard Business Review,* February 2007.

Dijksterhuis, Ap, Rick B. van Baaren, Karin C. A. Bongers, Maarten Bos, Matthijs L. van Leeuwen, and Andries van der Leij. "The Rational Unconscious: Conscious Thought versus Unconscious Thought in Complex Consumer Choice." Working paper.

Dijksterhuis, Ap, Maarten Bos, Loran Nordgren, and Rick van Baaren. "On Making the Right Choice: The Deliberation-Without-Attention Effect." *Science* 311 (2006): 1005–07.

Dijksterhuis, Ap, and Ad van Knippenberg. "The Relation Between Perception and Behavior, or How to Win a Game of Trivial Pursuit." *Journal of Personality and Social Psychology* 74 (1998): 865–77.

Dijksterhuis, Ap, and Loran Nordgren. "A Theory of Unconscious Thought." Working paper.

Dijksterhuis, Ap, and Zeger van Olden. "On the Benefits of Thinking Unconsciously: Unconscious Thought Can Increase Post-Choice Satisfaction." *Journal of Experimental Social Psychology* 42 (2006): 627–31.

Dolan, Ray, et al. "Dopaminergic Modulation of Impaired Cognitive Activation in the Anterior Cingulate in Schizophrenia." *Nature* 378 (1995): 180–82.

Dougherty, Pete, and Jim Wyatt. "Will Wonderlic Scores Cause Teams to Wonder about Young?" *USA Today*, March 1, 2006.

Dweck, Carol. *Mindset*. New York: Random House, 2006.

———. *Self-Theories*. Philadelphia: Psychology Press, 2000.

Ekman, Paul. "An Argument for Basic Emotions." *Cognition and Emotion* 6 (1992): 169–200.

El-Hai, Jack. *The Lobotomist*. New York: Wiley, 2007.

Fama, Eugene. "Random Walks in Stock Market Prices." *Financial Analysts Journal*, September/October 1965 (reprinted January/February 1995).

Fehr, Ernst, and Klaus Schmidt. "A Theory of Fairness, Competition, and Cooperation." *Quarterly Journal of Economics* 71 (1999): 397–404.

Finlan, Alastair. *The Royal Navy in the Falklands Conflict and the Gulf War*. London: Routledge, 2004.

Fleck, Jessica, and Robert Weisberg. "The Use of Verbal Protocols as Data: An Analysis of Insight in the Candle Problem." *Memory and Cognition* 32 (2004): 990–1006.

Fleming, Renee. *The Inner Voice*. New York: Viking, 2004.

Frank, Robert. *Passions Within Reason*. New York: Norton, 1988.

Freud, Sigmund. *Civilization and Its Discontents*. New York: Norton, 2005.

———. *New Introductory Lectures on Psycho-analysis*. London: Hogarth Press, 1933.

Gailliot, M. T., R. Baumeister, C. N. DeWall, J. K. Maner, E. A. Plant, D. M. Tice, L. E. Brewer, and B. J. Schmeichel. "Self-Control Relies on Glucose as a Limited Energy Source: Willpower Is More Than a Metaphor." *Journal of Personality and Social Psychology* 92 (2007): 325–36.

Galvan, Adriana, Todd A. Hare, Cindy E. Parra, Jackie Penn, Henning Voss, Gary Glover, and B. J. Casey. "Earlier Development of the Accumbens Relative to Orbitofrontal Cortex Might Underlie Risk-Taking Behavior in Adolescents." *Journal of Neuroscience* 25 (2006): 6885–92.

Gaspar, P., et al. "Catecholamine Innervation of the Human Cerebral Cortex as Revealed by Comparative Immunohistochemistry of Tyrosine Hydroxylase and Dopamine-$\beta$-hydroxylase." *Journal of Comparative Neurology* 279 (1988): 249–71.

Gazzaniga, Michael. *The Social Brain.* New York: Basic Books, 1985.

Gazzaniga, Michael, ed. *The New Cognitive Neurosciences.* Cambridge: MIT Press, 2006.

Geier, Andrew, Paul Rozin, and Gheorghe Doros. "Unit Bias: A New Heuristic that Helps Explain the Effect of Portion Size on Food Intake." *Psychological Science* 17 (2006): 521–27.

George, Carol, and Mary Main. "Social Interactions of Young, Abused Children." *Child Development* 50 (June 1979): 306–18.

Gigerenzer, Gerd. *Gut Feelings.* New York: Viking, 2007.

Gigerenzer, Gerd, and Reinhard Selten, eds. *Bounded Rationality: The Adaptive Toolbox.* Cambridge: MIT Press, 2000.

Gilbert, Daniel. *Stumbling on Happiness.* New York: Knopf, 2007.

Gilovich, Thomas. *How We Know What Isn't So: The Fallibility of Human Reason in Everyday Life.* New York: Free Press, 1991.

Gilovich, Thomas, Robert Vallone, and Amos Tversky. "The Hot Hand in Basketball: On the Misperception of Random Sequences." *Cognitive Psychology* 17 (1985): 295–314.

Gladwell, Malcolm. *Blink.* Boston: Little, Brown and Company, 2005.

———. "The Art of Failure." *The New Yorker,* August 21, 2000.

Glimcher, Paul. "Decisions, Decisions, Decisions: Choosing a Biological Science of Choice." *Neuron* 36 (2002): 323–32.

———. *Decisions, Uncertainty, and the Brain.* Cambridge: MIT Press, 2003.

Glimcher, Paul, and Aldo Rustichini. "Neuroeconomics: The Consilience of Brain and Decision." *Science* 306 (2004): 447–52.

Goodwin, Doris Kearns. *Team of Rivals.* New York: Simon and Schuster, 2005.

Gordon, Dan, ed. *Your Brain on Cubs.* New York: Dana Press, 2007.

Graden, Brian. Interviewed on "The Merchants of Cool," *Frontline,* PBS, February 26, 2001; http://www.pbs.org/wgbh/pages/frontline/shows/cool/interviews/graden.html.

Green, Donald, Bradley Palmquist, and Eric Schickler. *Partisan Hearts and Minds.* New Haven: Yale University Press, 2002.

Greene, Joshua, S. A. Morelli, K. Lowenberg, L. E. Nystrom, and J. D. Cohen. "Cognitive Load Selectively Interferes with Utilitarian Moral Judgment." *Cognition* 107 (2008): 1144–54.

Greene, Joshua, L. E. Nystrom, A. D. Engell, J. M. Darley, and J. D. Cohen. "The Neural Bases of Cognitive Conflict and Control in Moral Judgment." *Neuron* 44 (2004): 389–400.

Greene, Joshua, R. Brian Sommerville, Leigh E. Nystrom, John M. Darley, and Jonathan D. Cohen. "An fMRI Investigation of Emotional Engagement in Moral Judgment." *Science* 293 (2003): 2105–08.

Grossman, Dave. *On Combat: The Psychological Cost of Learning to Kill in War and Society.* Boston: Back Bay Books, 1996.

Grove, William, David Zald, Boyd Lebow, Beth Snitz, and Chad Nelson. "Clinical versus Mechanical Prediction: A Meta-Analysis." *Psychological Assessment* 12 (2000): 19–30.

Gucciardi, Daniel, and James A. Dimmock. "Choking under Pressure in Sensorimotor Skills: Conscious Processing or Depleted Attentional Resources?" *Psychology of Sport and Exercise* 9 (2008): 45–59.

Haidt, Jonathan. *The Happiness Hypothesis.* New York: Basic Books, 2006.

———. "The Emotional Dog and Its Rational Tail: A Social Intuitionist Approach to Moral Judgment." *Psychological Review* 108 (2001): 814–34.

Hayden, Benjamin, and Michael Platt. "Fool Me Once, Shame on Me —Fool Me Twice, Blame the ACC." *Nature Neuroscience* 9 (2006): 857–58.

Heilman, Kenneth. *Matter of Mind: A Neurologist's View of Brain-Behavior Relationships.* Oxford: Oxford University Press, 2002.

Helmreich, Robert. "Managing Human Error in Aviation." *Scientific American* 276 (May 1997): 62–67.

Henrich, Joseph, et al. *Foundations of Human Sociality: Economic Experiments and Ethnographic Evidence from Fifteen Small-Scale Societies.* Oxford: Oxford University Press, 2005.

Hoffman, Elizabeth, Kevin McCabe, and Vernon Smith. "Social Distance and Other-Regarding Behavior in Dictator Games." *The American Economic Review* 86 (1996): 653–60.

Hogarth, Robin. *Educating Intuition.* Chicago: University of Chicago Press, 2001.

Holden, Anthony. *Big Deal: A Year as a Professional Poker Player.* New York: Simon and Schuster, 2007.

Hollerman, Jeffrey, and Wolfram Schultz. "Dopamine Neurons Report an Error in the Temporal Prediction of Reward during Learning." *Nature Neuroscience* 1 (1998): 304–09.

Homer-Dixon, Thomas. *The Ingenuity Gap.* New York: Vintage, 2002.

Inman, J. Jeffrey, Leigh McAlister, and Wayne Hoyer. "Promotion Signal: Proxy for a Price Cut?" *Journal of Consumer Research* 17 (1990): 74–81.

Ito, S., et al. "Performance Monitoring by the Anterior Cingulate Cortex During Saccade Countermanding." *Science* 302 (2003): 120–22.

James, William. *The Principles of Psychology.* Vol. 2. New York: Dover, 1950.

Jarvik, Jeffrey, et al. "Rapid Magnetic Resonance Imaging versus Radiographs for Patients with Low Back Pain." *Journal of the American Medical Association* 289 (2003): 2810–18.

Jefferson, Thomas. *The Writings of Thomas Jefferson.* Washington, DC: Lipscomb and Bergh, 1903.

Jensen, Maureen, Michael N. Brant-Zawadzki, Nancy Obuchowski, Michael T. Modic, Dennis Malkasian, and Jeffrey S. Ross. "Magnetic Resonance Imaging of the Lumbar Spine in People without Back Pain." *New England Journal of Medicine* 331 (1994): 69–73.

Juckel, G., et al. "Dysfunction of Ventral Striatal Reward Prediction in Schizophrenia." *NeuroImage* 29 (2006): 409–16.

Jung-Beeman, Mark, et al. "Neural Activity Observed in People Solving Verbal Problems with Insight." *Public Library of Science—Biology* 2 (2004): 500–10.

Kahneman, Daniel, et al., eds. *Judgment under Uncertainty: Heuristics and Biases.* Cambridge: Cambridge University Press, 2000.

Kahneman, Daniel, and Amos Tversky. "Prospect Theory: An Analysis of Decision under Risk." *Econometrica* 47 (1979): 263–91.

Kahneman, Daniel, and Amos Tversky, eds. *Choices, Values, and Frames.* Cambridge: Cambridge University Press, 2002.

Kandel, Eric, et al. *Principles of Neural Science.* New York: McGraw Hill, 1999.

Keltner, Dacher. "The Power Paradox." *Greater Good* 4 (Winter 2007–2008).

Kennerley, Steven, et al. "Optimal Decision-Making and the ACC." *Nature Neuroscience* 9 (2006): 940–47.

Kermer, D. A., Erin Driver-Linn, Timothy Wilson, and Daniel Gilbert. "Loss Aversion Is an Affective Forecasting Error." *Psychological Science* 17 (2006): 649–53.

Klein, Gary. *The Power of Intuition.* New York: Doubleday, 2004.

———. *Sources of Power.* Cambridge: MIT Press, 1999.

Klein, L. R. "In the Digital Age: An Empirical Study of Prepurchase Search for Automobiles." *Journal of Interactive Marketing* 17 (2003): 29–49.

Klein, Tilmann, et al. "Genetically Determined Differences in Learning from Errors." *Science* 318 (2007): 1462–65.

Knutson, Brian, Scott Rick, G. Elliott Wimmer, Drazen Prelec, and George Loewenstein. "Neural Predictors of Purchases." *Neuron* 53 (2007): 147–56.

Koenigs, M., et al. "Damage to the Prefrontal Cortex Increases Utilitarian Moral Judgments." *Nature* 446 (2007): 908–11.

Kounios, John, J. L. Frymiare, E. M. Bowden, J. I. Fleck, K. Subramaniam, T. B. Parrish, and Mark Jung-Beeman. "The Prepared Mind: Neural Activity Prior to Problem Presentation Predicts Solution by Sudden Insight." *Psychological Science* 17 (2006): 882–90.

Lehrer, Jonah. "The Psychology of Back Pain." *Best Life*, February 2008.

Levenson, Robert, L. L. Carstensen, and J. M. Gottman. "The Influence of Age and Gender on Affect, Physiology, and Their Interrelations: A

Study of Long-Term Marriages." *Journal of Personality and Social Psychology* 67 (1994): 56–68.

Lo, Andrew, and Dmitry Repin. "The Psychophysiology of Real-Time Financial Risk Processing." *Journal of Cognitive Neuroscience* 14 (2002): 323–39.

Logothetis, N., and Josef Pfeuffer. "On the Nature of the BOLD fMRI Contrast Mechanism." *Magnetic Resonance Imaging* 22 (2004): 1517–31.

Lohrenz, Terry, Kevin McCabe, Colin Camerer, and P. Read Montague. "Neural Signature of Fictive Learning Signals in a Sequential Investment Task." *Proceedings of the National Academy of Sciences* 104 (2007): 9493–98.

MacDonald, Angus, Jonathan D. Cohen, V. Andrew Stenger, and Cameron S. Carter. "Dissociating the Role of the Dorsolateral Prefrontal and Anterior Cingulate Cortex in Cognitive Control." *Science* 288 (2000): 1835–39.

Mackay, Alan. *A Dictionary of Scientific Quotations*. New York: Taylor and Francis, 1991.

Maclean, Norman. *Young Men and Fire*. Chicago: University of Chicago Press, 1992.

Main, Mary, and Carol George. "Responses of Abused and Disadvantaged Toddlers to Distress in Agemates: A Study in the Day Care Setting." *Developmental Psychology* 21 (1985): 407–12.

Mangels, J. A., et al. "Why Do Beliefs about Intelligence Influence Learning Success? A Social-Cognitive-Neuroscience Model." *Social, Cognitive, and Affective Neuroscience* 1 (2006): 75–86.

Marcus, Gary. *Kluge*. Boston: Houghton Mifflin, 2008.

Marshall, S.L.A. *Men Against Fire: The Problem of Battle Command*. Tulsa: University of Oklahoma Press, 2000.

Masserman, Jules, Stanley Wechkin, and William Terris. "Altruistic Behavior in Rhesus Monkeys." *American Journal of Psychiatry* 121 (1964): 584–85.

McCabe, Kevin, Daniel Houser, Lee Ryan, Vernon Smith, and Theodore Trouard. "A Functional Imaging Study of Cooperation in Two-Person Reciprocal Exchange." *Proceedings of the National Academy of Sciences* 98 (2001): 11832–35.

McClure, Samuel, G. Berns, and P. Montague. "Temporal Prediction Errors in a Passive Learning Task Activate Human Striatum." *Neuron* 38 (2003): 339–46.

McClure, Samuel, David Laibson, George Loewenstein, and Jonathan Cohen. "Separate Neural Systems Value Immediate and Delayed Monetary Rewards." *Science* 306 (2004): 503–07.

McClure, Samuel, Jian Li, Damon Tomlin, Kim Cypert, Latané Montague, and P. Montague. "Neural Correlates of Behavioral Preference for Culturally Familiar Drinks." *Neuron* 44 (2004): 379–87.

McManus, James. *Positively Fifth Street.* New York: Picador, 2004.

Miller, E. K., and J. D. Cohen. "An Integrative Theory of Prefrontal Function." *Annual Reviews of Neuroscience* 24 (2001): 167–202.

Miller, George. "The Magical Number Seven, Plus or Minus Two: Some Limits on Our Capacity for Processing Information." *Psychological Review* 63 (1956): 81–97.

Mlodinow, Leonard. *The Drunkard's Walk.* New York: Pantheon, 2008.

Moll, J., Frank Krueger, Roland Zahn, and Matteo Pardini. "Human Fronto-mesolimbic Networks Guide Decisions about Charitable Donation." *Proceedings of the National Academy of Sciences* 103 (2006): 15623–28.

Montague, Read. "The First Wave." *Trends in Cognitive Sciences* 11 (2007): 407–09.

———. "Neuroeconomics: A View from Neuroscience." *Functional Neurology* 22 (2007): 219–34.

———. *Why Choose This Book?* New York: Dutton, 2006.

Montague, Read, Steven Hyman, and Jonathan Cohen. "Computational Roles for Dopamine in Behavioral Control." *Nature* (2004) 431: 760–67.

Montague, Read, Brooks King-Casas, and Jonathan Cohen. "Imaging Valuation Models in Human Choice." *Annual Review of Neuroscience* 29 (2006): 417–48.

Muller, S. B., et al. "How Do World-Class Cricket Batsmen Anticipate a Bowler's Intention?" *Quarterly Journal of Experimental Psychology* 59: 2162–86.

Myers, David. *Intuition.* New Haven: Yale University Press, 2004.

Napoli, Philip. *Audience Economics.* New York: Columbia University Press, 2003.

Odean, Terrance. "Are Investors Reluctant to Realize Their Losses?" *Journal of Finance* 53 (October 1998): 1775–98.

Olds, James, and Peter Milner. "Positive Reinforcement Produced by Electrical Stimulation of Septal Area and Other Regions of Rat Brain." *Journal of Comparative and Physiological Psychology* 47 (1954): 419–27.

Oosterbeek, Hessel, Randolph Sloof, and Gijs van de Kuilen. "Differences in Ultimatum Game Experiments: Evidence from a Meta-Analysis." *Experimental Economics* 7 (2004): 171–88.

Ovid. *Metamorphoses.* Book VII. Translated by D. Raeburn. London: Penguin, 2004.

Oya, H., et al. "Electrophysiological Correlates of Reward Prediction Error Recorded in the Human Prefrontal Cortex." *Proceedings of the National Academy of Sciences* 102 (2005): 8351–56.

Page, Scott. *The Difference*. Princeton: Princeton University Press, 2008.

Parker, Susan, et al. "The Impact of Early Institutional Rearing on the Ability to Discriminate Facial Expressions of Emotion: An Event-Related Potential Study." *Child Development* 76 (2005): 54–72.

Pierce, Charles. *Moving the Chains*. New York: Farrar, Straus and Giroux, 2007.

Plassmann, Hilke, John O'Doherty, Baba Shiv, and Antonio Rangel. "Marketing Actions Can Modulate Neural Representations of Experienced Pleasantness." *Proceedings of the National Academy of Sciences* 105 (2007): 1050–54.

Plato. *Phaedrus*. Translated by Alexander Nehamas and Paul Woodruff. New York: Hackett, 1995.

Post, Thierry, et al. "Deal or No Deal? Decision Making under Risk in a Large-Payoff Game Show." *American Economic Review*, March 2008.

Predmore, Steven. "The Dynamics of Group Performance: A Multiple Case Study of Effective and Ineffective Flightcrew Performance." PhD dissertation, University of Texas at Austin, 1992.

Prelec, Drazen, and Duncan Simester. "Always Leave Home Without It." *Marketing Letters* 12 (2001): 5–12.

Prinz, Jesse. *Gut Reactions*. New York: Oxford University Press, 2004.

Rabinovich, Abraham. *The Yom Kippur War: The Epic Encounter That Transformed the Middle East*. New York: Schocken, 2005.

Raviv, Dan, and Yossi Melman. *Every Spy a Prince*. Boston: Houghton Mifflin, 1990.

Rolls, Edmund. *The Brain and Emotion*. New York: Oxford University Press, 1999.

———. "The Orbitofrontal Cortex and Reward." *Cerebral Cortex* 10 (2000): 285–94.

Rorie, A. E., and W. T. Newsome. "A General Mechanism for Decision-Making in the Human Brain?" *Trends in Cognitive Science* 9 (2005): 41–43.

Sandkuhler, Simone, and Joydeep Bhattacharya. "Deconstructing Insight: EEG Correlates of Insightful Problem Solving." *PLoS One*, January 2008, http://www.plosone.org/article/fetchArticle.action?articleURI =info:doi/10.1371/journal.pone.0001459.

Sanfey, Alan, James K. Rilling, Jessica A. Aronson, Leigh E. Nystrom, and Jonathan D. Cohen. "The Neural Basis of Economic Decision Making in the Ultimatum Game." *Science* 300 (2003): 1755–57.

Schelling, Thomas. *Choice and Consequence*. Cambridge: Harvard University Press, 1985.

Schoenemann, P. T., M. J. Sheehan, and L. D. Glotzer. "Prefrontal White Matter Volume Is Disproportionately Larger in Humans Than in Other Primates." *Nature Neuroscience* 8 (2005): 242–52.

Schultz, Robert, et al. "The Role of the Fusiform Face Area in Social Cognition: Implications for the Pathobiology of Autism." *Philosophical Transactions of the Royal Society, Series B* 358 (2003): 415–27.

Schultz, Wolfram. "Predictive Reward Signal of Dopamine Neurons." *Journal of Neurophysiology* 80 (1998): 1–27.

Schultz, Wolfram, P. Dayan, and P. R. Montague. "A Neural Substrate of Prediction and Reward." *Science* 275 (1997): 1593–99.

Schwartz, Barry. *The Paradox of Choice.* New York: HarperCollins, 2004.

Shaw, Philip, Jason Lerch, Deanna Greenstein, Wendy Sharp, Liv Clasen, Alan Evans, Jay Giedd, F. Xavier Castellanos, and Judith Rapoport. "Attention-Deficit/Hyperactivity Disorder Is Characterized by a Delay in Cortical Maturation." *Proceedings of the National Academy of Sciences* 104 (2007): 19649–54.

Shiv, Baba, Ziv Carmon, and Dan Ariely. "Placebo Effects of Marketing Actions: Consumers May Get What They Pay For." *Journal of Marketing Research* (November 2005): 383–93.

Shiv, Baba, and Alexander Fedorikhin. "Heart and Mind in Conflict: The Interplay of Affect and Cognition in Consumer Decision Making." *Journal of Consumer Research* 26 (1999): 278–92.

Shiv, Baba, George Loewenstein, Antoine Bechara, Hanna Damasio, and Antonio Damasio. "Investment Behavior and the Negative Side of Emotion." *Psychological Science* 16 (2005): 435–39.

Shoda, Y., W. Mischel, and P. K. Peake. "Predicting Adolescent Cognitive and Self-regulatory Competencies from Preschool Delay of Gratification." *Developmental Psychology* 26 (1990): 978–86.

Small, D. A., George Loewenstein, and Paul Slovic. "Sympathy and Callousness: The Impact of Deliberative Thought on Donations to Identifiable and Statistical Victims." *Organizational Behavior and Human Decision Processes* 102 (2007): 143–53.

Smith, Adam. *The Theory of Moral Sentiments.* New York: Prometheus Books, 2000.

Sniderman, Paul, Richard Brody, and Philip Tetlock. *Reasoning and Choice: Explorations in Political Psychology.* Cambridge: Cambridge University Press, 1993.

Solomon, Robert. *True to Our Feelings.* New York: Oxford University Press, 2006.

Sommers, Tamler. "Jonathan Haidt." *The Believer,* August 2005.

Steele, Claude. "Thin Ice: Stereotype Threat and Black College Students." *Atlantic Monthly,* August 1999.

Steele, Claude, and J. Aronson. "Stereotype Threat and the Intellectual Test Performance of African-Americans." *Journal of Personality and Social Psychology* 69 (1995): 797–811.

Stout, Martha. *The Sociopath Next Door.* New York: Broadway, 2005.

Sullivan, Terry, and Peter Maiken. *Killer Clown: The John Wayne Gacy Murders.* New York: Pinnacle, 2000.

Sutton, Rich. "Learning to Predict by the Methods of Temporal Difference." *Machine Learning* 3 (1998): 9–44.

Tankersley, D., C. J. Stowe, and S. A. Huettel. "Altruism Is Associated with an Increased Neural Response to Agency." *Nature Neuroscience* 10 (2007): 150–51.

Tavris, Carol, and Elliot Aronson. *Mistakes Were Made (But Not by Me).* New York: Harvest, 2008.

Taylor, S. E., N. I. Eisenberger, D. Saxbe, B. J. Lehman, and M. D. Lieberman. "Neural Bases of Regulatory Deficits Associated with Childhood Family Stress." *Biological Psychiatry* 60 (2006): 269–301.

Telfer, Ross, and John Biggs. *The Psychology of Flight Training.* Des Moines: Iowa State Press, 1988.

Tesauro, Gerald, et al. "A Parallel Network That Learns to Play Backgammon." *Artificial Intelligence* 39 (1989): 357–90.

Tetlock, Philip. *Expert Political Judgment.* Princeton: Princeton University Press, 2006.

Thaler, Richard. *The Winner's Curse.* Princeton: Princeton University Press, 1992.

Thaler, Richard, and Cass Sunstein. *Nudge: Improving Decisions and Health, Wealth and Happiness.* New Haven: Yale University Press, 2008.

Trivers, Robert. "The Evolution of Reciprocal Altruism." *Quarterly Review of Biology* 46 (1971): 35–67.

Tversky, Amos, and Daniel Kahneman. "Judgment under Uncertainty: Heuristics and Biases." *Science* 185 (1974): 1124–31.

———. "The Framing of Decisions and the Psychology of Choice." *Science* 211 (1981): 453–58.

Vallone, Robert, et al. "The Hostile Media Phenomenon: Biased Perception and Perceptions of Media Bias in Coverage of the Beirut Massacre." *Journal of Personality and Social Psychology* 49 (1985): 577–85.

Wager, Tor, et al. "Placebo-Induced Changes in fMRI in the Anticipation and Experience of Pain." *Science* 303 (2004): 1162–66.

Wansink, Brian. *Mindless Eating.* New York: Bantam, 2006.

Weltman, G., J. E. Smith, and G. H. Egstrom. "Perceptual Narrowing During Simulated Pressure-Chamber Exposure." *Human Factors* 13 (1971): 99–107.

Westen, Drew. *The Political Brain.* New York: Public Affairs, 2007.

Westen, Drew, C. Kilts, P. Blagov, K. Harenski, and S. Hamann. "The Neural Basis of Motivated Reasoning: An fMRI Study of Emotional Constraints on Political Judgment during the U.S. Presidential

Election of 2004." *Journal of Cognitive Neuroscience* 18 (2006): 1947–58.

Wilkinson, Alec. "Conversations with a Killer." *The New Yorker*, April 18, 1994.

Wilson, M., and M. Daly. "Do Pretty Women Inspire Men to Discount the Future?" *Proceedings of the Royal Society of London Biological Science* 51: 1326–35.

Wilson, Timothy. *Strangers to Ourselves*. Cambridge: Harvard University Press, 2002.

Wilson, Timothy, and Daniel Gilbert. "Affective Forecasting." *Advances in Experimental Psychology* 35: 345–411.

Wilson, Timothy, Douglas Lisle, Jonathan Schooler, Sara D. Hodges, Kristen J. Klaaren, and Suzanne J. LaFleur. "Introspecting About Reasons Can Reduce Post-Choice Satisfaction." *Personality and Social Psychology Bulletin* 19 (1993): 331–39.

Wilson, Timothy, and Jonathan Schooler. "Thinking Too Much: Introspection Can Reduce the Quality of Preferences and Decisions." *Journal of Personality and Social Psychology* 60 (1991): 181–92.

Wood, Jacqueline, and Jordan Grafman. "Human Prefrontal Cortex: Processing and Representational Perspectives." *Nature Reviews Neuroscience* 4 (2003): 139–50.

Wynn, Thomas, and Frederick Coolidge. "The Expert Neandertal Mind." *Journal of Human Evolution* 46 (2004): 467–87.

Zillgitt, Jeff. "Shape Up or Bust Out." *USA Today*, May 30, 2008.

Zweig, Jason. *Your Money and Your Brain*. New York: Simon and Schuster, 2008.

# Index